Praise for *Where the Water Goes*

"It's a rare writer who can explain the inexplicable, but David Owen manages to do just that. *Where the Water Goes* is at once informative, entertaining, and unsparing—essential reading for anyone who cares about the American West."

—Elizabeth Kolbert, author of *The Sixth Extinction*

"Fascinating, thoughtful, and wise. David Owen is an extraordinarily gifted writer."

—Bill Bryson, author of *The Road to Little Dribbling* and *A Walk in the Woods*

"Owen has the keen observation of a birder combined with the breezy writing to draw you in with unusual insights. . . . As Owen shows, the Colorado River is a great, sad, terrifying, possibly hopeful example of the pervasive, permanent mark people are making on the planet." —*The New York Times Book Review*

"This wonderfully written book covers issues that will, or should, give you a headache. But it is a good headache, one that makes you a more informed person. Mr. Owen writes about water, but in these polarized times the lessons he shares spill into other arenas. The world of water rights and wrongs along the Colorado River offers hope for other problems." —*The Wall Street Journal*

"Owen is effortlessly engaging, informally parceling out information about acre-foot allotments alongside sketches of notable, often dreadful figures in the river's history. . . . *Where the Water Goes* doesn't pretend to solve the problems Owen acknowledges are overwhelming and, in some ways, impossible. It's a restless travelogue of long-term human impact on the natural world, and how politics and economics have as much to do with redirecting rivers as any canal. But with its historical eddies, policy asides, and trips to the Hoover Dam, at heart *Where the Water Goes* is about water as a function of time, and a reminder that we're running out of both." —NPR.org

"*Where the Water Goes* makes an eloquent argument for addressing the impact of human inhabitants on the natural world." —BBC.com

"Part road-trip documentary, part memoir, and part geopolitical and hydrology lesson, author David Owen's book follows the historical and geographic course of the river, the water it carries, and the lives that depend on it. . . . [Owen] effectively describes the links between historic precedents, choices, and events that led the river and the millions of people who depend upon it to the present state." —*Science*

"The story Owen tells in *Where the Water Goes* is crucial to our future."

—*Boulder Daily Camera*

T0004951

"David Owen's new book, *Where the Water Goes: Life and Death Along the Colorado River* . . . handles its sprawling subject with deftness and quirkiness. . . . Owen delves into the history and politics of the much-dammed, over-allocated river, as well as the arcana of Western water law and the weirdness of RV culture, without losing sight of larger questions about the sustainability of America's efforts to make the desert bloom."
—*Westword*

"This gorgeous new book is a compelling and fascinating read about the Colorado River, a crucial water source for a surprisingly large portion of the United States. David Owen, with utmost elegance and wry wit, examines the river from headwaters to terminus and all the stops along the way."
—BookPeople blog

"[A] revealing investigation of hydroecology in extremis . . . Rather than simply bemoan environmental degradation, Owen presents a deeper, more useful analysis of the subtle interplay between natural and human needs."
—*Publishers Weekly*

"An essential read for not only the environmentally minded but also citizens who are curious about where their water comes from. Highly recommended."
—*Library Journal*

"Owen offers a wealth of engrossing and often surprising details about the complicated nature of water rights, recreational usage (worth $26 billion a year), and depletion threats from climate change and the fracking industry. With water shortages looming across the globe, Owen's work provides invaluable lessons on the rewards and pitfalls involved in managing an essential natural resource."
—*Booklist*

"An important work that brings the questions surrounding water use in the American Southwest forward to the era of climate change. With humor, an acute eye, and unshowy skill, Owen has written a book that deserves to stand with Marc Reisner's classic, *Cadillac Desert*."
—Ian Frazier, author of *Great Plains, On the Rez,* and *Hogs Wild*

"I have traveled the American West all of my life and thought that I knew everything about its fabled water wars. But David Owen fills in so many gaps that I feel that I've been to water reeducation camp. Whether you read for fun, or edification, this is a gem."
—Rinker Buck, author of *The Oregon Trail: A New American Journey*

WHERE
THE
WATER
GOES

LIFE AND DEATH ALONG
THE COLORADO RIVER

DAVID OWEN

RIVERHEAD BOOKS
New York

RIVERHEAD BOOKS
An imprint of Penguin Random House LLC
375 Hudson Street
New York, New York 10014

Copyright © 2017 by David Owen

Parts of several chapters of this book first appeared, in different form, in *The New Yorker.*
Some material about Las Vegas first appeared, in different form, in *Golf Digest.*

The Library of Congress has catalogued the Riverhead hardcover edition as follows:

Names: Owen, David, date.
Title: Where the water goes : life and death along the Colorado River / David Owen.
Description: New York : Riverhead Books, [2017] |
Includes bibliographical references and index.
Identifiers: LCCN 2016039410 | ISBN 9781594633775 |
ISBN 9780698189904 (e-ISBN)
Subjects: LCSH: Colorado River (Colo.–Mexico)—Description and travel. |
Colorado River (Colo.–Mexico)—Environmental conditions. | Owen, David—
Travel—Colorado River (Colo.–Mexico) | Water-supply—West (U.S.) |
Stream ecology—Colorado River (Colo.–Mexico)
Classification: LCC F788 .O84 2017 | DDC 917.91/304—dc23
LC record available at https://lccn.loc.gov/2016039410
p.   cm.

First Riverhead hardcover edition: April 2017
First Riverhead trade paperback edition: April 2018
Riverhead trade paperback ISBN: 9780735216099

Printed in the United States of America
ScoutAutomatedPrintCode

Book design by Meighan Cavanaugh

Map by Sarah Evans Lloyd. Reference: Pacific Institute, Oakland, California

For Alice and Hugh O'Keefe

CONTENTS

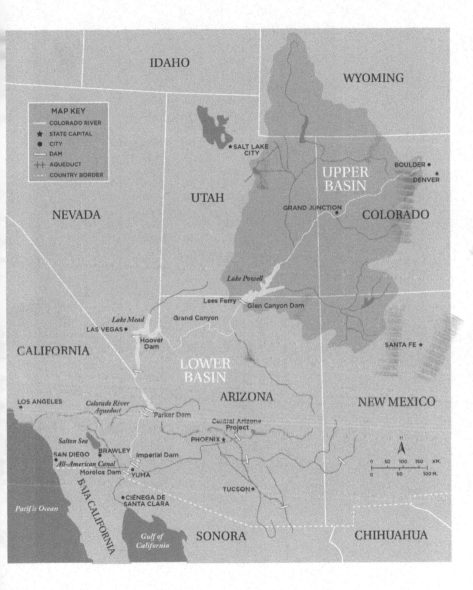

1.

THE HEADWATERS

Our pilot, David Kunkel, asked me to retrieve his oxygen bottle from under my seat, and when I handed it to him he gripped the plastic breathing tube with his teeth and opened the valve. We had taken off from Boulder not long before and were flying over Rocky Mountain National Park, thirty miles to the northwest. Kunkel was navigating with the help of an iPad Mini, which was resting on his legs. "People don't usually think altitude is affecting them," he said. "But if you ask them to count backward from a hundred by sevens they have trouble." What struck me at that moment was not how high we were but how low: a little earlier, we had flown within what seemed like hailing distance of the sheer east face of Longs Peak, and now, as Kunkel banked steeply to the right to give us a better view of a stream at the bottom of a narrow valley, his wingtip appeared to pass just feet from the jagged declivity beneath us. Snow had fallen in the mountains during the night, and I half expected it to swirl up in our wake.

The other passenger, sitting in the copilot's seat and leaning out the window with a big camera, was Jennifer Pitt, who at the time was a senior researcher for the Environmental Defense Fund. Pitt is in her forties. She has long brown hair, which she had pulled back into a ponytail, and she was wearing a purple fleece. She worked at the EDF, mostly on issues related to the Colorado River, from 1999 till 2015, when she moved to a similar job at the National Audubon Society. In recent years, her focus has been on the river's other end, in Mexico, but she had agreed to show me its source. Our principal destination that day was the Colorado's headwaters, just over the Continental Divide, roughly fifty miles south of the Wyoming state line. "The best way to see a river system is from the air," she'd told me earlier. She arranged our flight through LightHawk, an international nonprofit organization that supplies volunteer pilots and their airplanes, at no charge, for a variety of environmental purposes. The previous day, a LightHawk pilot had flown twenty black-footed ferrets from Fort Collins to a spot near the Grand Canyon, for relocation.

Before our flight, I looked up Kunkel on Google and was disconcerted to find a news story about him landing his Cessna 340 on a highway in the Rockies after losing both engines in succession. But then I realized that nothing like that could happen to us, because the plane he'd be using for our trip, a Maule M-7, had just one engine. I looked up Pitt, too. She was born in Boston and grew up in Westchester County, New York, in a suburb of New York City. "I think you can trace my interest in rivers back to my childhood in Westchester," she told me later, "because I grew up in a river town, on the Hudson, and when I was a kid Pete Seeger came to my school and sang to me about rivers." As an undergraduate, at Harvard, she majored in American history and literature, but developed an interest in urban planning and landscape architecture. "After graduation," she continued, "I worked in Manhattan for a year, for the Department of Parks and Recreation, and realized that that was not what I wanted to do." She got a job as an interpretive ranger in Mesa Verde National Park, in southwestern Colorado, and that experience, she said, "gave another twist to my view of the world, and how an ancient culture used the resources around them." She earned a master's degree in environmental sciences, with a focus on water, at the Yale School of Forestry, then worked in Washington, D.C., for five years, mostly at the National Park Service. In 1999, the Environmental Defense Fund hired her to create programs related to the Colorado River and the ecosystems that depend on it. In 2003, she married Michael Cohen, a senior associate at the Pacific Institute, another environmental organization. (They met at a water conference in Tucson.) They live in Boulder and have a daughter.

Kunkel dipped a wing, and Pitt pointed toward the Never Summer Mountains, on our right. "There's the Grand Ditch," she said. I saw what looked like a road or a hiking trail cut across the face of a steeply sloping forest of snow-dusted conifers; she explained that it was an aqueduct, dating to 1890. Its original full name was the North Grand River Ditch. (Until 1921, the section of the Colorado that's upstream from its confluence with the Green, in eastern Utah, was called the Grand. Hence: Grand Lake, Grand Junction, Grand Valley—but not Grand Canyon, which was named for its grandness.) It was built with pickaxes and black powder, mostly by Japanese laborers, and it operates by gravity—an impressive feat of pre-laser engineering. The Grand Ditch is fourteen miles long, and much of it is above ten thousand feet. It carries water across the Continental Divide at La Poudre Pass and empties it into a stream that flows toward the state's eastern plains, where even by the late 1800s farmers were feeling parched. It doesn't tap the Colorado directly, but captures as much as forty percent of the flow from slopes that would otherwise feed it, like a sap-gathering gash in the trunk of a rubber tree. We had already flown over several larger, more recent additions to the same network: Long Draw Reservoir, completed in 1930; Estes Lake, which serves as a trans-basin junction box; and five connected natural and man-made lakes that lie on the western side of the divide and gather and store water from the Colorado or its watershed. The northernmost of the lakes spills as much as a third of a billion gallons a day into the Alva B. Adams Tunnel, which was built in the 1940s. Adams was a lawyer and a U.S. senator, and in the early 1930s he served as the chairman of the Committee on Irrigation and Reclamation. The tunnel moves the water under the center of the park, drops it through five hydroelectric generating plants, and delivers it to a distribution system that serves a populous area east of the mountains, including Boulder. The main elements of the system are known collectively as the Colorado-Big Thompson Project. (In the West, "project" almost always means "dam," "reservoir," "aqueduct," "canal," or all four.)

Kunkel made a slow turn to the left. "We just flew over the headwaters," he said. Our position was easier to see on his iPad than on the ground. The sky had been blue when we took off, but since we'd entered the mountains he'd had to pick his way under and around what sometimes looked like an upside-down ocean of clouds. The ceiling made flying difficult but helped to explain the existence of the water-storing-and-shifting network we'd been looking at. As moisture-laden weather systems move eastward across the western United States, they pile up over the Rockies, dumping snow and rain. Eighty percent of Colorado's precipitation falls on the western half of the state, yet eighty-five percent of the population lives to the east, in the mountains' "rain shadow." If transporting water from one side to the other were impossible, most of the people who live and farm on the eastern side of the mountains would have to move. Pitt said, "Even people who describe themselves as worried environmentalists usually have no idea where their water comes from. We did a focus group once where somebody asserted vehemently that Denver did not get any water from the other side of the mountains, and we actually had to intervene and make sure that the guy leading the focus group knew that that was wrong, so that the whole two-hour discussion didn't go off in some other direction."

. . .

WHEN THE FIRST EUROPEANS to see the Grand Canyon looked down from its southern rim, in 1540, they guessed that the stream they could see at the bottom must be about eight feet wide. They'd been fooled by the scale of the canyon, but even so the Colorado River isn't huge. Marc Reisner, in *Cadillac Desert*, his classic book about water in the western United States, published in 1986, calls it "comparatively miniature." If you were to spread a full year's worth of its entire flow evenly over a surface the size of its drainage basin, roughly 250,000 square miles, the water would cover it to a depth of only about an inch. Forty years ago, I crossed the Mississippi River on a tiny ferry, now long gone, which operated between northwestern Tennessee and southeastern Missouri. My car and I were the only cargo, and as we churned toward the setting sun I took swigs from a bottle of whiskey, which I'd bought in Kentucky a couple of hours earlier at the House of Bourbon, a liquor store, and by the time we got to the opposite bank the sun was almost gone and I was a little bit drunk. No part of the Colorado is so wide that crossing it would give you time to become even slightly buzzed, except in places where dams have turned the river into lakes. The Mississippi is also a thousand miles longer. It carries the equivalent of the Colorado's entire annual flow every couple of weeks.

Yet the Colorado is a crucial resource for a surprisingly large part of the United States. In a report published in 1861, James Christmas Ives, an Army lieutenant who had led a mapping expedition up the river from its mouth in northern Mexico a few years before, described the Grand Canyon as "altogether valueless," and concluded: "It seems intended by nature that the Colorado River, along the greater portion of its lonely and majestic way, shall be forever unvisited and undisturbed." It didn't remain that way for long. Less than seventy years later, a congressman described the Colorado, more accurately, as "intrinsically the most valuable stream in the world." It and its tributaries flow through or alongside seven western states—Colorado, Wyoming, Utah, New Mexico, Nevada, Arizona, and California—before crossing into Mexico near Yuma, Arizona. It supplies water to more than 36 million people, including residents not just of Boulder, Denver, and Colorado Springs but also of Salt Lake City, Albuquerque, Las Vegas, Phoenix, Tucson, San Diego, and Los Angeles, several of which are hundreds of miles from its banks. It irrigates close to six million acres of farmland, much of which it also created, through eons of silt deposition. It powers two of the country's largest hydroelectric plants, at Hoover Dam and Glen Canyon Dam, as well as many smaller ones. It's the principal water source for two enormous man-made reservoirs, Lake Mead and Lake Powell, as well as many smaller ones. It supports recreational activities that are said to be worth $26 billion a year. Some of its southern reaches attract so many transient residents during the winter that you could almost believe it had overflowed its banks and left dense alluvial deposits of motorboats, Jet Skis, dirt bikes, golf carts, all-terrain vehicles, RVs, mobile homes, fifth wheels, and people with gray hair.

All that human utility has costs. The Colorado has helped to shape some of the most otherworldly landforms on earth—the Grand Canyon, of course, and also the Vermilion Cliffs, in northern Arizona, and the eerily striated buttes and mazelike sandstone meanders of Canyonlands National Park, in southeastern Utah—yet even within those seemingly wild landscapes its flow is so altered and controlled that in many ways the river functions more like a fourteen-hundred-mile-long canal. The legal right to use every gallon is owned or claimed by someone—in fact, more than every gallon, since theoretical rights to the Colorado's flow, known to water lawyers as "paper water," greatly exceed its actual flow, known as "wet water." That imbalance has been exacerbated by the drought in the western United States, which began just before the turn of the millennium, but even if the drought ended tomorrow, problems would remain. The river has been "over-allocated" since the states in its drainage basin first began to divide the water among themselves, nearly a century ago.

The Colorado suffers from the same kinds of overuse and environmental degradation that increasingly threaten freshwater sources all over the world, as the global population rises toward its projected mid-century level of nine or ten billion, and as changes in the weather play havoc with accustomed precipitation patterns. ("Climate change is water change," a scientist told me.) Water challenges in the United States are less dire than those in places like India, Syria, and Brazil, but they're similar in kind. They also involve much more than water, since they're inextricable from equally thorny challenges concerning energy, economics, governance, democracy, and climate.

Water problems are straightforward in one way: without water we die, and not centuries from now. When supplies are short, people have no choice but to find solutions, one way or another, in real time. They change behavior, cut back consumption, develop new sources, negotiate treaties, pass legislation—all right now—and we know that happens because in dry places all over the world there's evidence of it every day. Water problems in the western United States, when viewed from afar, can seem tantalizingly easy to solve: all we need to do is turn off the fountains at the Bellagio, stop selling hay to China, ban golf, cut down the almond trees, and kill all the lawyers. As you draw closer, though, you realize that every new solution creates additional problems and that tinkering with even small elements in the river's vast network of beneficiaries can upset dozens of others. Addressing everything effectively, equitably, and permanently will force us to weigh the kinds of choices we prefer to avoid. We haven't had much success with that sort of thing in the past. But who knows?

. . .

I GREW UP IN KANSAS CITY, and, like many people who grow up in Kansas City, I spent a lot of time thinking about places far away. Most of all, I thought about the Rocky Mountains, six hundred miles to the west. I first visited Colorado on vacation with my family when I was six, and I returned many times: for summer camp, for adventures with friends, for backpacking, rock climbing, and skiing, and for my freshman and sophomore years of college. The first big purchase I made with money that I myself had earned was a fancy mountaineering sleeping bag. I hung it on a wall in my bedroom at home, both to keep the goose down from compacting and to remind my parents that I was just passing through.

In 1976, when I was twenty-one, I spent the summer living in a rented house in Colorado Springs and working on the grounds crew of an apartment complex on what was then the outskirts of the city. During most of the week, my coworkers and I moved hoses and sprinklers around the property, to keep the grass green, and then we mowed what we had grown. We had to be at work at six-thirty but weren't allowed to begin mowing until seven, to allow the residents to sleep in, so for the first half hour we devised alternative methods of waking them up, such as slowly dragging shovels across parking lots. Watering was like a race. The grass began to turn brown almost the moment we moved our sprinklers, partly because we were a mile above sea level in what is essentially a desert, and partly because the apartment complex had been built on porous ground, on the site of an old quarry. One night, I dreamed that one of the Rain Bird rotary sprinklers we used at work was keeping me awake by rhythmically spraying me in bed, and I made a mental note to ask my housemate not to water my room while I was trying to sleep.

Among the many questions I failed to ask myself that summer was where all the water we used at work came from. All I knew was that every time I attached a hose to a spigot and turned it on, I could run it full force until it was time to go home. I now know that the city's water in those days came from local surface streams and wells. I also know that, since then, the Colorado Springs metropolitan area has more than doubled in population and sprawled far into the Eastern Plains, and that today much of its water comes from the other side of the mountains. The most important source is the Fryingpan-Arkansas Project, a vast man-made water-moving network a hundred miles or so southwest of Colorado-Big Thompson. It collects snowmelt from the watershed of the Fryingpan River—a tributary of the Roaring Fork, which is a tributary of the Colorado—and moves it under the Continental Divide to users in the state's southeastern quadrant. The Fry-Ark consists of six big dams, sixteen small dams, 4 miles of canals, 27 miles of tunnels, and 282 miles of conduits. It was authorized by President Kennedy in 1962 and took almost twenty years to complete, and the average user of the water it delivers doesn't know that it exists.

I learned about the Fry-Ark not long ago, while following the Colorado River from beginning to end. I had decided that a useful way to think about water issues of all kinds would be to trace the course of a single river, to see where the water came from and where it went. The Colorado is an ideal subject for such a study, both because of its economic importance—it has been called "the American Nile"—and because at fourteen hundred miles it's short enough to allow a traveler to trace its course but long enough to cross a great deal of varied terrain. Following the Colorado also gave me an opportunity to wander around in a part of the country where I once believed I was destined to spend my adult life (living in a cabin, climbing mountains, making pemmican, eating plants I'd read about in *Stalking the Wild Asparagus*, and writing poems in a battered notebook).

I didn't travel the river in a boat, the way John Wesley Powell did; I followed it on land, in a succession of rental cars. During several week-long trips, the first of which began two days after my flight with Pitt and Kunkel, I explored as much of the Colorado as I could without getting wet. I drove more than three thousand miles; made many stops, detours, and redundant loops; and listened to three of the five volumes of the audiobook of *Game of Thrones*. I also received what I now think of as a graduate-level education in the river and its many dependents, human and otherwise.

The Colorado provides an especially useful introduction to water issues because we literally use it up. The river's historical outlet is at the northern end of the Gulf of California, also known as the Sea of Cortez, where the Baja Peninsula joins the mainland like an arm attaching itself to a torso. But people who depend on the Colorado divert so much water as the river winds through the southwestern United States that since the early 1960s it has seldom flowed all the way to the end, and since the late 1990s has made it only once. There's a point, not far from the border, where the water simply runs out, and from there to the gulf what ought to be the river's streambed becomes difficult to distinguish from the arid expanse on either side. For most of the past fifty years, the Colorado's historical delta, which once was a complex and intermittently verdant wetland, has been a million-acre desert. People who drive into or out of the town of San Luis Río Colorado, in the Mexican state of Sonora, sometimes complain about having to pay a six-peso toll to cross a bridge that spans only sand.

I BEGAN MY JOURNEY, more or less by accident, very close to the river's headwaters. I'd been interviewing someone in Eagle, Colorado, 130 miles west of Denver, and I had a dinner appointment that evening back in Boulder, so I typed the address of the restaurant into the Google navigation app on my phone and was surprised but pleased to see that the suggested route ignored I-70, which runs right past Eagle, and took me instead up into the mountains on minor roads and then alongside the Colorado almost all the way to its source before crossing the divide and descending into Boulder from the west. A couple of weeks earlier, back at home, a different Google app had offered to navigate me to "work." I hadn't known what to make of that, because my office is in my house, so I clicked the tab and discovered that Google had deduced, based on how I spend my time during a typical week, that I must work at 10 Golf Course Road—the address of my golf club. More recently, I'd made many Web searches related to the Colorado River; had Google noticed those, too, and tailored my Boulder route to suit my obvious interest?

The route began on a two-lane highway. Then the road stopped having painted lines; then it stopped having pavement; then it shrank to the width of a driveway. I got out of my car and looked down at a long coal train passing far below me, on tracks that followed the twisting stream. The train ran through a short tunnel and along a skinny shelf that had been carved into a talus slope above the far bank, perhaps a hundred feet higher than the water. Much of the Colorado is too narrow, too shallow, too rugged, too plunging, and too full of big rocks ever to have been used for commercial navigation, the way the Mississippi always has been, but, for as long as humans have lived or traveled nearby, its valleys and canyons have served as transportation corridors through what would otherwise be impassable terrain.

The canyon I was looking at was named for Sir St. George Gore, an Anglo-Irish baronet, who passed through on horseback in 1854 on his way farther west. Gore was the absentee owner of seven thousand acres in Ireland. In the late 1840s, during the Great Famine, he had evicted tenants who couldn't pay their rents and sent them off to North America in ships overloaded with the starving, and he was so despised by the tenants who remained that in 1849 they murdered his agent. An informational sign I'd seen earlier said that he and his party, during their western trip, had "slaughtered 2,000 buffalo, 1,600 elk and deer, and 100 bears." Gore, who was known as the Buffalo Slayer, did almost all the killing himself. He also shot many thousands of mountain sheep, coyotes, wolves, and birds. He was indifferent to the suffering of animals he had merely wounded, and he left the vast majority of the carcasses to rot. His luggage included a brass bed, a steel bathtub decorated with his coat of arms, a fur-covered commode, French carpets, 112 horses, eighteen oxen, forty mules, four dozen hunting dogs, six large wagons, twenty-one carts, seventy-five rifles, many shotguns and pistols, and three tons of ammunition. Among the members of his traveling party were a team of taxidermists and a man whose only job was tying flies. The legendary mountain man Jim Bridger served as one of his guides. The U.S. secretary of the interior denounced Gore's slaughter, which he said threatened the food supplies of several Indian tribes, but the government took no action. William "Buffalo Bill" Cody called Gore "a sportsman among a thousand." Gore returned to Ireland in 1857, the year Lieutenant Ives began his expedition up the river from its other end. He visited the United States once more, in 1874, but this time his destination was Florida, where he focused on ridding the Everglades of alligators, egrets, and flamingos.

The ribbon of water at the bottom of the Gore Canyon looked sinuous and beautiful and cold—it was greenish-gray and churning—but the road I was following wasn't much of a road, and it kept shrinking. Then, near Kremmling, I noticed that my phone didn't say I was one hour and four minutes from Boulder, as I'd thought it did; it said I was one day and four hours. And suddenly I understood that when I'd entered the address of my destination, back in Eagle, I must have accidentally clicked the "pedestrian" icon on my navigation app: Google thought I was on foot. Getting to Boulder from that spot by car was somewhat complicated, but I didn't regret my mistake, because the complete trip took me alongside streams and lakes and reservoirs and diversions that I'd also seen from Kunkel's airplane, as well as ones that I had not. Pitt was right when she said the best way to study a river system is from above, but some of its features need to be seen from the ground, too. The audacity of the Grand Ditch didn't fully strike me until I'd crossed the divide myself, not far from La Poudre Pass, and imagined the project's evolution in the minds of the nineteenth-century dreamers who conceived it. They built their aqueduct without bulldozers, modern explosives, or help from the Global Positioning System, and they did it all so well that, more than a century later, the water still flows. The distances alone are mind-boggling. How did people who really were on foot decide that moving water across a mountain range was even a possibility?

My journey along the Colorado took me to farms, government offices, campgrounds, power plants, ghost towns, fracking sites, aqueducts, reservoirs, and pumping stations, and it gave me opportunities to lose myself in some truly jaw-dropping topography. My journey ended in Mexico, in a truck belonging to someone else. In that truck, a Mexican environmentalist drove Jennifer Pitt and me across an expanse of sand to a point where the river ceased to exist. Where had the water gone? By then, I had a pretty good idea.

2.

THE LAW OF THE RIVER

My river trip actually began the day before I met Jennifer Pitt, and it started with paper water, rather than wet. I flew to Colorado from New York, and spent part of an afternoon with Kent Holsinger, a lawyer whose several specialties include western water law. We met in a conference room at the headquarters of one of his clients, the Colorado Oil & Gas Association, in downtown Denver. (He's the chairman of COGA's water committee.) He was wearing a checked shirt—pretty much the uniform of the western American male—and he had a tiny bit of gray in his goatee, which was otherwise light brown. He grew up on a cattle ranch near Walden, Colorado, a town that, on a clearer day, Kunkel, Pitt, and I could have spotted from the air. His parents still lived there.

"It's a small cattle ranch, about eight hundred acres," he told me. "My parents have kind of slowed down, so we don't run our own cattle anymore, but we lease out the pasture in the summer, and we put up irrigated native-grass hay, which we sell primarily to horse people." A stream crosses the ranch, and the Holsingers draw water from it, but their right to do so isn't based on the fact that their property is adjacent to its banks, as it would be in the East. "Water law in Colorado and most states in the West is based on the doctrine of 'prior appropriation,'" he said. That doctrine holds that the first person to make "beneficial use" of water gains the right to use that quantity for that purpose forever, and that the claim takes precedence over every claim made later and is unrelated to the user's distance from the stream.

The prior-appropriation doctrine originated during the California gold rush, which began in 1849, and during a similar gold rush in Colorado a decade later. The most common mining technique in that era involved shoveling earth into a wooden trough, called a "sluice box," and then directing a swiftly moving stream of water over it. Gold is dense, and bits of it will usually settle into riffles in the bottom of a sluice box and remain there after the lighter material has been washed away. Successful sluicing depends on access to lots of water, and deadly disputes arose when newcomers made diversions upstream from existing operations. In the East and in England, most surface water was (and is) governed by "riparian law," whose guiding principle is that the right to draw water from a stream must be shared equitably by all adjacent property owners. That didn't work with gold, because in the West there was so little water that dividing a stream among multiple users often made it useless to all. Early farmers—more than a few of whom were former miners who had given up on mining—faced identical conflicts when they tried to irrigate.

The solution, through much of the West, was a new conception of water allocation, whose central tenet was "first in time, first in right." Proximity to the source counted for nothing—because miners often had to move water long distances to reach new mining claims, and arable land wasn't always close to streams. (A Colorado farmer's right to move water across land belonging to other people was formally established in 1872; the right to move water out of one drainage basin for use in another was affirmed ten years later, in *Coffin v. Left Hand Ditch*

Company.) Under the new system, the critical factor in all water disputes was the date of first use. Holsinger told me that his family's water right is more than a century old, but that a downstream neighbor's is older, and that the neighbor therefore had the legal right to draw his full allotment before his parents draw theirs. "If the senior wasn't getting all his water, he would call the water-rights commissioner," Holsinger said. "The commissioner would go down the list to the next junior, and if that was us he would make the dreaded call, and say we had to turn off our headgate"—a valve that diverts water from a stream, in their case to irrigate pastures and hay fields. "We've got a mile of river on the property, but that didn't make any difference. Our right was junior, so whenever the river was low we had to shut down, and that meant turning off our income. That's why senior water rights are extraordinarily valuable."

In Colorado, water rights are governed by a separate court system, established in 1969; it has seven jurisdictions, one in each of the state's seven main river drainages. No water right is official until a water court has "adjudicated" it by issuing a "decree," a sort of legal declaration of authenticity. The decree certifies the appropriation date, which establishes the owners' place in line, and it states the quantity of water covered by the right, as well as the purpose for which the water may legally be used. A decreed water right is "conditional" until the water is actually put to its decreed use, at which point (assuming the court agrees) it becomes "perfected." Owners of conditional rights are allowed time in which to perfect their claim but are required to demonstrate diligence in doing so—by beginning the construction of an irrigation ditch, say, or by entering into lawsuits with environmental groups trying to stop them—and if they don't demonstrate diligence they risk losing their "reserved" place on the priority list. Owners of water rights cannot sell them, or modify their decreed use, without the approval of the court, and, if they seek that approval, potentially affected users are notified and given an opportunity to object.

It's all actually much more complicated than that. There's a very good book, *Colorado Water Law for Non-Lawyers* by P. Andrew Jones and Tom Cech, but parts of it are hard to understand unless you're the sort of non-lawyer who happens to have gone to law school. At any rate, probably the most significant feature of the prior-appropriation system is that, during times of shortage, sacrifices are made from the bottom of the priority list up, rather than shared. The website of Colorado's Division of Water Resources offers the hypothetical example of a stream supplying water to three users with adjudicated rights, the first and second most senior of whom hold decrees for two cubic feet per second (cfs), and the most junior of whom holds a decree for one cubic foot per second. "When the stream is carrying 5 cfs or more, all of the rights on the stream can be fulfilled," the website explains. "However, if the stream is carrying only 3 cfs of water, its priority number 3 will not receive any water, with priority number 2 receiving only half of its 2 cfs right." Neighbors don't always adhere to the rules; David Kunkel, the pilot on Jennifer Pitt's and my flight over the headwaters, used to own a ranch in the northwestern part of the state. (He was selling it at the time of our trip.) The ranch was in the watershed of a stream outside the drainage of the Colorado, and, Kunkel said, "Nobody ever imposed a call on it"—meaning that when water was scarce the rights holders had always worked things out among themselves, without drying up those who happened to be "out of priority." And in times of severe shortage even large senior users have sometimes agreed to be accommodating. But in most places, at most times, allocations are made in accordance with the priority list, and senior rights holders are under no obligation to share.

"A great early example of prior appropriation is Horace Greeley's

Union Colony," Holsinger said. Greeley was the editor of the *New-York Tribune*. He was a financial backer of the Union Colony of Colorado, also known as the Union Temperance Colony, a utopian community situated where the city of Greeley is today. He promoted western migration ("Go West, young man!") and was among those responsible for spreading the scientifically baseless idea that "rain follows the plow," which held that human habitation and crop cultivation generated precipitation, transforming formerly arid western microclimates into arable Edens—an idea made semi-credible by record rainfall in the late 1860s and again in the late 1880s. Cynthia Barnett, in *Rain: A Natural and Cultural History*, published in 2015, writes, "Some Americans were convinced that rail and telegraph lines expanding to the West were triggering violent thunderstorms. The financier Jay Gould, at a time when he controlled both the Union Pacific and Missouri Pacific railroad companies and Western Union Telegraph, made the claim that railroad and telegraph construction was expanding the nation's rainy district by some twenty miles west a year." The problem was compounded by the fact that land grants made in the West under the Homestead Act of 1862 were too small to sustain commercially viable livestock grazing, even with irrigation. (Each homestead grant was a "quarter section"— one-fourth square mile, 160 acres—a considerable spread in the East, but just a fifth the size of Kent Holsinger's parents' "small cattle ranch.") Various combinations of those factors were eventually responsible for the abandonment of many farms and ranches, and the ruination of many lives.

Most of the water used by the Union Colony came not from rainfall but from the South Platte River, a resource so unreliable that its streambed, during dry periods, also served as a wagon trail. "The colony's residents established a system of canals to irrigate their land," Holsinger said, "but during a drought in 1874 irrigators in Fort Collins, which is upstream, literally dried them up." The colony's farmers complained, arguing that they had built their irrigation network first, and the two communities eventually reached an agreement that acknowledged Union's priority. (Parts of the original Union ditch system are still in use.) Soon afterward, arguments and lawsuits arising from similar conflicts in the same region gave the first-in-time principle the force of law. "All this was happening as Colorado was writing its constitution," Holsinger said, "and the prior-appropriation system was incorporated into it." Because of that, and because Colorado, among all the western states, employs what may be the strictest interpretation of the basic concept, prior appropriation is sometimes referred to as the Colorado Doctrine.

MOST OF THE WATER in the Colorado River originates in snowpack in mountains in the northern part of its watershed, but the biggest consumers of that water are at the river's far end—in southern California especially. In the early 1900s, people in the other river states, where settlement was still thin, worried that California was growing so quickly that its farmers and municipal water systems would establish priority claims to virtually all the water, even though the Colorado doesn't actually flow through California or receive water from tributaries originating there, but forms its border with Arizona. In 1922, representatives from the seven states that touch either the river or the rivers that feed it—Wyoming, Colorado, Utah, New Mexico, Arizona, Nevada, and California—met in Santa Fe, New Mexico, to work out a sharing agreement that would supersede or, at any rate, supplement the strict requirements of the prior-appropriation doctrine. The negotiations had begun in Washington eleven months earlier and had nearly collapsed several times, usually over disagreements about how much water each state would be entitled to draw. Colorado, which receives most of the precipitation that feeds the river, argued that the division should be based on precipitation; Arizona, which contains tributaries that feed the Colorado, felt strongly that tributaries shouldn't count against a state's allotment; California, the oldest big user, could find no flaw in the first-in-time principle.

The chairman of the meetings was Herbert Hoover, who was President Warren Harding's secretary of commerce. The breakthrough compromise, proposed by him and the delegate from Colorado, was to defer the issue of individual shares, by splitting the river's watershed into two imaginary basins, upper and lower, and dividing its natural flow equally between them. The upper-basin states, by the terms of the agreement, are Wyoming, Colorado, Utah, and New Mexico; the lower-basin states are Nevada, Arizona, and California. The dividing line was drawn at Lees Ferry, an isolated outpost in north central Arizona. Each basin was granted an average of 7.5 million acre-feet per year. (An acre-foot is the amount of water that would cover an acre to a depth of a foot—roughly 326,000 gallons, or about 1,230 cubic meters. Various rules of thumb place average U.S. household consumption, including both indoor and outdoor uses, at anywhere from one to three families per acre-foot per year—although consumption varies widely. An Olympic-size swimming pool contains almost exactly two acre-feet.) An additional two or three million acre-feet was left without a definite allocation; some of that amount, the delegates agreed, would probably eventually go to Mexico, the northern part of which is crossed by the Colorado, if the United States, "as a matter of international comity," ever decided that Mexico had "any right to the use of any waters of the Colorado River System." (Mexico was mentioned only because Hoover insisted.) Working out the many unresolved details, including state-by-state shares within the basins, took several more decades and required the involvement of Congress, the National Guard, the Supreme Court, and Henry Kissinger. But the basic agreement, which is known as the Colorado River Compact, is still in force.

The authors of the compact sidestepped or ignored many issues, including the water needs of American Indian tribes, a number of which occupy huge reservations within the Colorado's watershed. And they made no reference to anything that anyone today would identify as a concern for the natural environment. In the 1920s, "conserving" river water meant extracting as much profit from it as possible before it flowed into the sea. The fact that a natural resource might have value for species other than our own, or even that it might have aesthetic, spiritual, or recreational value for us, was not a consideration. Environmental threats were viewed mainly the other way around, as threats posed by the environment to people. Much of the impetus for building Hoover Dam, which was completed in 1936, came from downstream farmers who were less interested in storing water or generating electricity than in being protected from floods and inundations of silt, both of which occurred frequently and unpredictably when the Colorado was unconstrained.

THE COMPACT'S NEGOTIATORS viewed their division of the water as conservative, because hydrologists estimated that the river's typical "virgin flow" was at least 17 million acre-feet a year—more than enough to satisfy the agreement while leaving what everyone at the time believed to be lots of room for growth, even for California. But subsequent studies, including tree-ring analyses, have proven that the hydrologists were wrong. It's now known that the years on which the original estimates were based, in the early twentieth century, were the wettest since the 1400s, and that 1922, the year of the agreement, was one of the wettest of all. The authors of a law journal article published in 1986 wrote that "this mistake of fact was so phenomenal as to appear more like a trick of fate." Since the 1920s, there have been years when the total flow was less than a third of what the negotiators assumed it would be, and scientists have identified ancient dry periods that lasted many decades—periods known as "megadroughts."

For most of the twentieth century, the errors made by hydrologists a century ago were inconsequential, because they affected only paper water: people hadn't yet found ways to routinely use more than all the wet water in the river. (The upper-basin states still take less than their theoretical entitlement; in 2012, they used only about sixty percent of it.) Even today, the impact on human activity has been less than it might have been, because the river's two huge reservoirs, Lake Mead and Lake Powell, have been treated like lower-basin credit cards. In 1998, both lakes were full and, between them, stored more than 50 million acre-feet—roughly three and a half years' worth of the river's average total flow. Today, they contain much less than half that amount, mainly because the lower-basin states have drawn them down by taking water that the compact and other agreements have allowed them to take. In a paper published in 2008, two scientists at the Scripps Institution of Oceanography wrote that "currently scheduled depletions are simply not sustainable." According to the Government Accountability Office, total consumption of water from the Colorado River system in 1985—when Colorado River users still worried more about floods than about drought—amounted to 28.7 million acre-feet. That's roughly double what's now believed to be the long-term average annual flow of the river; it's also roughly 6.5 million acre-feet more than is currently stored in Mead and Powell combined.

Climate change isn't the only source of uncertainty, but it's the biggest one, since its predicted likely effects include significantly decreased snowfall in the mountains that feed the river. Barton H. "Buzz" Thompson, who teaches natural resources law at Stanford Law School and was appointed by the Supreme Court in 2008 as the special master in a water-rights dispute between Montana and Wyoming, told me, "If you look at paleo-climate data, you see that, roughly during the medieval era, there were some very significant dry periods in the Colorado's basin. It wasn't necessarily drought every year; there were times when there might be five or six years of drought, then a couple of somewhat wetter years. But, overall, things kept getting drier and drier. And some of those extended dry periods lasted between something like 70 and 140 years. Temperatures at those times tended to be higher than the average over the past two thousand years—but the differences were nowhere near as great as the temperature increase we've experienced over the past twenty or thirty years. So it raises concerns."

It also raises a potentially monumental legal issue. Article III of the compact says that the upper-basin states "will not cause the flow of the river at Lee Ferry to be depleted below an aggregate of 75,000,000 acre-feet for any ten consecutive years." But what if depletion is caused not by the direct actions of the upper-basin states but by long-term changes in upper basin precipitation patterns? The question, Jennifer Pitt told me, is "whether you are held harmless because you didn't de-

‣—climate change did." That's a legal issue that has never been

᠁rt, and Pitt said there are many theories about what would

-"which, hopefully, it never will."

bove, the allocations specified by the compact

d; in fact, in 1944 the agreed normal annual

5 million acre-feet to accommodate Mexico. At

ᷧe Colorado River Water Users Association in

aesars Palace in Las Vegas, two representatives

clamation—the federal agency that oversees

—estimated that the system now has an aver-

.2 million acre-feet a year. I ran into Pitt on

ᷧting (which she attends every year). She told

ᷧr almost certainly understated the imbal-

ficant was that the deficit was being openly

anted—something, she said, that hadn't

been true in the past, when the shortages were easier to ignore and most water managers treated environmentalists as impertinent nuisances.

The prior-appropriation doctrine and the Colorado River Compact are central elements of what's known throughout the Colorado's watershed as the Law of the River, a complex but loosely defined and minimally circumscribed body of rules, precedents, habits, treaties, customs, and compacts that isn't written down all in one place but is invoked almost any time two water users disagree about who's entitled to what. Grady Gammage, Jr., a lawyer and real estate developer in Phoenix, once told an interviewer that, when he first became involved in water issues, he felt that every time he made a comment about the Colorado another lawyer would inform him that whatever he had just suggested was "prohibited by the Law of the River." Gammage had been in practice for some time, but didn't recognize the reference. "So I go to the Arizona Revised Statues book and pull it down, and I look up 'River, comma, Law of,' and it's not there." He did the same with the United States Code, also without success. "It turns out that the Law of the River is kind of like the British constitution," he continued. "It's whatever the people who have really been hanging around it a long time think it is." Invoking it, furthermore, is a privilege reserved for those who ha undergone what Gammage called "the Water Buffalo ceremonia mittance initiation rites." Water Buffaloes are old-school western experts: managers, engineers, diverters, legislators, and lawyers all of them men, whose long immersion in river-related discu guments, negotiations, and lawsuits has made them deep of non–Water Buffaloes and has convinced them that w many ways, less significant than paper water. After year the compact and its successors, Gammage said, he ha Buffalo, too: "I've graduated to the point where I c would violate the Law of the River.'"

3.

TRIBUTARIES

The Colorado is more than a river. It's the main element, but not the only element, of a vast and intricately interconnected system, which draws water from tributaries of all sizes, and forms a drainage basin that covers a significant fraction of the western United States. Many people whose lives or incomes depend on the Colorado divert their water not from the river itself but from streams that feed it. Among the easternmost of those streams is the Blue River, which arises near Quandary Peak, a few miles from the ski-resort town of Breckenridge, Colorado. The Blue flows roughly sixty miles north-northwest and empties into the Colorado near Kremmling, not far from the spot where, during my drive from Eagle to Boulder, I realized that Google Maps thought I was walking.

The flow of the Blue River, like the flow of nearly all of the Colorado's principal tributaries, is not unimpeded. In the early 1960s, Denver Water, the state's largest water utility, dammed it in a mountain valley a little over an hour west of the city, near what's now the town of Silverthorne. The dam is an enormous, sloping, earth-filled wall more than a mile long and 230 feet high. Lake Dillon, the reservoir it created, covers thirty-two hundred acres, has a capacity of a quarter-million acre-feet, and supplies forty percent of the water used by Denver Water's customers. And for many people who live west of the Rockies it's a source of aggravation and annoyance.

On my first day on the road, I stopped for the night in Silverthorne and got a room at La Quinta Inn & Suites, just off I-70. From my window, on the sixth floor, I could look directly across a narrow valley to the dam, which looms above both the highway and the town, less than a thousand yards from the motel. Silverthorne began as a campsite for dam workers. The tunnel that carries water from the lake under the Continental Divide is ten feet in diameter and a little over twenty-three miles long. Digging it took six years, during which mining crews worked on it continuously, sometimes in twelve-hour shifts, and mainly used pickaxes, jackhammers, and dynamite. They dug in four directions at once: from both ends toward the center, and from the center toward both ends. The crews working from the center reached their starting point, under the town of Montezuma, by descending a thousand-foot vertical shaft, which they had to dig first. The tunnel was named for Harold D. Roberts, the lawyer who established the city's legal right to build the dam and take the water. It empties into the North Fork of the South Platte River, and the river carries the water the rest of the way east.

Lake Dillon sits on top of what used to be the town of Dillon, which was named for a mid-nineteenth-century prospector. It started as a trading post and changed locations twice before 1900, mainly to maintain direct contact with the railroad, which was still moving around. Denver Water, anticipating the city's future needs, began buying houses in Dillon during the Great Depression, as residents went broke and lost their homes for failing to pay their property taxes; it made its final purchases shortly before it began building the dam. Denver Water also bought what it described in a newsletter in 1963 as "an attractive site,

among the fir and evergreen," on ranchland about a mile away, so that
the town's former residents could relocate there if they wanted to. A
cemetery, a church, a hydroelectric plant, and thirteen miles of highway had to be moved, too. The resurrected town of Dillon is across the
interstate from Silverthorne on high ground at the northeastern corner
of the lake. It has a population of about nine hundred—more than it
did before it was inundated. Remnants of the old Dillon still exist on
the bottom of the reservoir, two hundred feet down.

If you want to get people who live west of the Continental Divide
worked up about people who live east of the Continental Divide, Lake
Dillon is a good topic to start with. "One of the biggest banes of our
existence is water being diverted to the Front Range," a woman who
lives about fifty miles from the reservoir told me. In the view of many
West Slope residents, the damming of the Blue River and the drowning
of Dillon were unconscionable acts of mid-twentieth-century urban hu-
bris. Nevertheless, this issue, like all water issues, is not straightforward.
Lake Dillon, in addition to supplying water to the bathrooms, kitchens,
lawns, and golf courses of metropolitan Denver, is one of the foundations of the economy of the stunningly beautiful mountain region that
surrounds it. Most of that economy is based on tourism, and most of
the money that sustains it comes from people who live east of the mountains or in other states. Several of Colorado's most popular ski resorts—
Arapahoe Basin, Breckenridge, Copper Mountain, Keystone—are
within ten miles of the dam, and Vail is just a few miles farther. Visitors
and residents are drawn to the area not only by the resorts but also by
the year-round recreational enticements of Lake Dillon.

I HEADED WEST FROM SILVERTHORNE, then left I-70 a few miles past
Vail and followed Highway 24 south, up the valley of the Eagle River,
another tributary of the Colorado. I passed Mount of the Holy Cross

and Mount Elbert, both of which I had climbed during backpacking trips when I attended summer camp in the 1960s. The Holy Cross climb was especially memorable, because just after we reached the summit ridge the temperature plunged and we were hit by a sudden rain-snow-and-hail storm. Everyone's hair froze into a helmet of ice, and for a while getting back to our campsite seemed as though it might turn out to be impossible, and when we finally did get back we found that most of our tents had blown over—a late-summer version of the sort of intense high-elevation precipitation on which western rivers depend.

A few miles beyond Leadville, I turned west onto Route 82 and skirted the northern shore of Twin Lakes, a double reservoir whose original iteration was created in the late 1800s by damming the outlet of a natural glacial pool. Twin Lakes is a minor element of the Fryingpan-Arkansas Project—the trans-mountain diversion I described in the first chapter—which carries water from the Colorado River system across the mountains to the southeastern part of the state. Beyond Twin Lakes, Route 82 winds up through the mountains to Independence Pass, elevation 12,095 feet. During a car trip in early June 2011, my wife and I stopped at the top and asked a stranger to photograph us standing in deep snow a few feet from the roadway. The pass had only just reopened, after being closed by heavy snows since the previous November. Mountain snowpack can be thought of as a vast natural reservoir. The snow in the pass had fallen the winter before, and as it melted it was gradually feeding streams on both sides of the Continental Divide. The depth of the winter snowpack in places like Independence Pass can affect water supplies as far away as Mexico.

When I drove over the pass from Silverthorne, I saw much less snow. It was early October, so the previous winter's accumulation had mostly melted, and the coming winter's accumulation hadn't fallen yet. In another month, though, the state highway department would close the entire road until roughly Memorial Day, when crews using enormous plows would push a path through the snow again. From the summit, I descended steeply into a mountain valley, at the bottom of which Route 82 became the main street of Aspen. Aspen began as a mining town and today serves as a reservoir primarily of skiers, private jets, and modest-looking houses with seven-digit sales prices. Much of the town is crisscrossed by the remnants of an irrigation-ditch system that dates back to the late 1800s, when Aspen was a center for silver mining; the surviving ditches are the little streams you see running along the edges of sidewalks and yards. People who live in or visit Aspen today don't typically think of themselves as water diverters, but the lawns and the leafy parks and the tall trees that shade most of the town's residential areas wouldn't be there if the ditch system didn't exist. (The trees are especially thirsty.)

I took Route 82 through town and all the way up the valley to Carbondale, and then turned south on Route 133 and followed it along the Crystal River—which, like the Fryingpan, is a tributary of the Roaring Fork and, therefore, a tributary once removed of the Colorado. That night, I stayed at the Redstone Inn, sixteen miles up the Crystal Valley from Carbondale. The inn was built in 1902 as a dormitory for bachelor employees of the Colorado Fuel & Iron Company, which mined high-grade coal in seams near the river and turned it into coke in a long double row of beehive ovens. (Coke is to coal what charcoal is to firewood, approximately.) The ovens, at their peak, produced almost six million tons a year, and they still exist, on the other side of the road and the river, directly across from the inn's entrance.

CF&I's founder and president, John Cleveland Osgood, was one of the country's richest industrialists. He was also a believer in what a historical website describes as "a perplexing mix of feudalism, capitalism and industrial paternalism." On two little streets next to the bachelor dormitory he built eighty-four chalet-style cottages for employees who were married. He also built a school, a bathhouse, and a social club, whose amenities included a library, a theater, and a saloon. For himself and his second wife, a Swedish countess, he built Cleveholm Manor, a forty-two-room, 24,000-square-foot house on a seventy-two-acre parcel a mile away. The house, a gamekeeper's lodge, the inn, the chalet village, and the coking ovens are all listed, separately, in the National Register of Historic Places. The inn is open year-round, and I highly recommend it, even though my room was a little ragged and the food wasn't great.

THE CRYSTAL VALLEY upstream from Redstone used to be busier than it is today. Ute Indians lived there until prospectors, trappers, ranchers, farmers, and miners forcibly removed them. A small town, which no longer exists, arose to support the settlers, and especially the miners, and for a long time it was the final stop on the Crystal River Railroad, which mainly hauled coal over tracks laid parallel to the river. Marble as white as lard was discovered in the mountains above the valley in the late 1870s. A quarry near what became the town of Marble supplied the raw material for the Tomb of the Unknown Soldier, which was carved from a single block, and the exterior of the Lincoln Memorial; today, most of its customers are in Europe and Asia. The gunfighter Doc Holliday lived in a borrowed cabin in the valley for a few months in 1887, futilely seeking relief from tuberculosis; he died that same year in a bed in a hotel room in Glenwood Springs, thirty miles down the river, and his last words were "Well, I'll be damned. This is funny." (He hadn't expected to die in bed, with his boots off.)

The railroad tracks were taken up in 1942 and recycled as war matériel, and most of the Crystal Valley between Redstone and the river's headwaters is currently relatively unmolested—a rarity among river basins of any size in Colorado. Since the 1980s, a coalition of conservationists has sought to have that section of the river designated "Wild and Scenic," in accordance with a 1968 act of Congress that extends federal protections to rivers that flow freely, possess certain "Outstandingly Remarkable Values," and meet other requirements. (The only river in the state with that designation currently is the section of the Cache la Poudre west of Fort Collins.) The original impetus for seeking protected status was a proposal, in the 1980s, to build two large dams, one a few miles above Redstone Inn and the other a little more than two miles below it. That proposal was a revived and modified version of the West Divide Project, which Congress approved in the 1960s but never funded. A few years ago, the Colorado River Water Conservation District, a state agency, tried again, this time with a proposal for a dam, a reservoir, and a power plant a short distance upstream from the inn. Mary Harris and Delia Malone were part of the group that defeated that proposal. They are both officers of the Roaring Fork chapter of the Audubon Society, and Malone is an ecologist and a research associate in the Colorado Natural Heritage Program of Colorado State University. They had agreed to show me where the dam was supposed to have gone.

We met in the parking lot of the Redstone Inn. Harris was wearing a maroon-and-black down jacket, and Malone was wearing a black down vest and gray yoga pants. (The morning was chilly.) Malone had brought her dog, a black Labrador retriever, and she had tied two strips of orange plastic to its collar, to protect it from hunters. We all got into Harris's SUV and drove up the valley—and it's a good thing we didn't take my car, I realized later, because when I started it the CD player would have picked up at an especially embarrassing moment in *Game of Thrones*, in the middle of the wedding night of Daenerys Targaryen and Khal Drogo. Arranged on a blue rubber mat on Harris's dashboard was a grouping of toy-soldier-size figures: angry farmers or townspeople from an unspecifiable foreign country and era, perhaps the ones who gathered with torches outside Castle Frankenstein and cursed Boris Karloff. The male figures were wearing hats and the female ones were wearing headscarves, and each was holding a weapon: a cleaver, a rolling pin, a rake, a pitchfork, a long knife, a rifle. Crouched near their feet were two green winged creatures, which looked a little like the flying monkeys in *The Wizard of Oz*. There was also an out-of-scale Uncle Duke, the Hunter S. Thompson character from the comic strip *Doonesbury*; he was wearing a T-shirt that said "Death Before Unconsciousness." Thompson, who died in 2005, lived in Woody Creek, a few miles away. The display represented Harris's exasperation with environmental desecraters.

A couple of miles upriver from the inn, tall rock walls pinched in on both sides of the road, and a little ways beyond that opening the valley broadened into an enormous mountain meadow, within which the Crystal had spread out into a maze of streams and ponds and grassy islands and meanders. We parked at a turnout and walked down a path to the edge of the water. On the far side of the valley, near the top of a dead spruce tree, I saw a pair of great blue heron nests, one above the other: huge, messy-looking saucers of silvery sticks. Harris thought she saw a bald eagle—Malone had seen one flying near the same spot the day before—but after passing binoculars back and forth we decided that the eagle was actually just sunlight glinting off a leafless branch. We did see an American dipper and a belted kingfisher, and, as we stood on the bank looking at other birds, the dog sat at Malone's feet and seemed to look, too.

We saw many signs of beavers, including a couple of lodges they'd erected beside broad meanders on the far side of the valley. Indeed, the entire section of river before us, as well as the meadow it flowed through, had been shaped by beavers over many generations. Dam-building is instinctive in beavers and may be triggered by flowing water, or the sound of flowing water. A similarly irresistible urge seems to be inborn in hydraulic engineers. I could easily picture the architects of the West Divide Project standing where we were standing and mentally adding up acre-feet. "It's the perfect place for a reservoir," Malone said. "They wouldn't have had to do much. That gap we drove through is the ideal spot for a dam." The valley floor even looked a little like the bottom of a lake.

ONE IMPETUS for the most recent revival of plans to dam the Crystal was a low-flow period a few years ago during which a local irrigator, by exercising his water right, temporarily dried up part of the river. To get the water to which his decree entitled him, the irrigator called out what's known as an "instream-flow" water right, which was junior to his own right. Instream-flow rights were invented by the state legislature in 1973, as a partial acknowledgment of the then still novel idea that protecting ecosystems might be a good thing. An instream-flow water right is like a water right that belongs to the river itself. (There's also a version for lakes.) The idea is to treat fish, other stream-dependent animals, and functioning wetlands as "beneficial uses," by assigning to some of the water they depend on its own place in a river's priority list. Instream-flow rights are limited in scope, however, and only the Colorado Water Conservation Board can appropriate them, and, because the idea of preserving the environment hasn't been around for as long as mining and farming and ranching have, instream flows usually have very junior priority dates—which means that during dry periods, when exercising them would be the most useful to whatever river system they're connected with, they inevitably get called out by something senior. The CWCB has the power to acquire water rights with older priority dates by buying or leasing them from willing sellers or lenders, but it doesn't have the power to acquire them by eminent domain. That's enough about western water law for now.

The proposed dam, in addition to providing irrigation water to farmers in another part of the state, would have enabled the local water district to bank water during the spring, when the Crystal River often floods, and release it later in the year, when the river is often low. That's the flood-control argument for reservoirs—one that downstream farmers in Arizona and California also made for what became Hoover Dam, back in the 1930s. There are a number of counterarguments, among them the fact that when you "control" flooding by damming rivers you disable a force by which riverine ecosystems maintain themselves. A paradoxical effect of the proposed dam, Malone and Harris said, is that it would have inundated a preexisting flood-control system with a long history of success: the beaver meadow we were looking at. "The beaver is the best thing the West ever had," Malone said. An area that would have been flooded by earlier versions of the West Divide Project is now a two-hundred-acre nature preserve, Filoha Meadows, about two miles downstream from the Redstone Inn. It has beaver dams, too, and a management plan for the preserve which was adopted in 2008 argued that the beavers there helped to support water levels during dry seasons. "Springtime overbanking flows are essentially stored in the beaver ponds and recharge groundwater," the plan said. "This water is then slowly discharged later when river flows decrease, ultimately helping maintain instream base flows long after snowmelt season has passed." Beavers also create habitats for other animals.

In fighting the most recent dam proposal, Harris, Malone, and the other members of their coalition had to go up against state and regional Water Buffaloes—the river veterans who shake their heads and cite the Law of the River whenever people like Harris and Malone disagree with them. "It was definitely a group of good-old-boy men," Malone told me. "They kept saying, 'Well, we can make the reservoir smaller.' But we said, 'No, we're not looking for a compromise.' They would just stare at us—like, 'damned women.'" Still, the women and their allies won, at least for the time being. (A representative of the river district told me later that the dam wasn't necessarily permanently dead.)

We got back in Harris's car, and, at a turnout a couple of miles away, we came across sixteen elementary school children on a field trip. I recognized one of the students as the daughter of the waitress who had taken my breakfast order that morning at the Redstone Inn. (She had been sitting at the bar and waiting for her school bus.) The students had pencils and spiral notebooks, and several were carrying backpacks, and most were exhibiting what I would consider a healthy amount of fidgetiness and boredom. One of their teachers asked them, "Do you know what it means to shoot yourself in the foot?" So we assumed they were talking about the dam.

GO WEST

From Redstone, I followed the valleys of the Crystal and Roaring Fork rivers to Glenwood Springs, then looped back eighteen miles to the east, to Dotsero, population seven hundred. If you're traveling in the other direction on I-70, Dotsero is the spot where you first encounter the Colorado River. (You also encounter the lava flow of the state's most recent volcanic eruption, which occurred more than four thousand years ago and left a crater thirteen hundred feet deep.) Dotsero began as a railroad junction during the 1800s, and, according to a popular and possibly true story, it was named inadvertently by a surveyor in the 1930s who made it the starting point—"dot zero"—of something he was drawing on a map. Roughly forty miles back up the Colorado from Dotsero, on the road to Kremmling, is a railroad junction that used to be called Orestod—the same name spelled backward.

At Dotsero, I reversed course and followed both I-70 and the river back to the west, through two remarkable canyons, Glenwood and De Beque, parts of which are so deep and narrow that some stretches of the roadway are engineered like double-decker bridges: one lane on top of the other. The railroad goes that way, too, on the opposite bank. Tunnels punch through buttresses of rock that the highway builders couldn't go around, and there are sections where the view above is so transfixing that I had to remind myself to look back at the road. I first stuck a hand in the river at the Bair Ranch rest area in Glenwood Canyon, six miles west of Dotsero. The rest area was named for an old sheep ranch. The Colorado is about two hundred feet wide at that point, and the water was cold and moving fast. An older couple in an RV had stopped, too, and the three of us looked thoughtfully at the river and read the informational signs mounted to the railings. They started at the leftmost sign and I started at the rightmost, and when we crossed in the center they told me they were on their way to the national parks in southern Utah. A few miles farther west, I passed the Shoshone Hydroelectric Plant, a century-old electricity generating facility. The plant is so close to both the river and I-70 that I actually drove over part of it. The facility—a cluster of barnlike tan buildings—is easy to miss; what I mainly noticed was a pair of huge, pale-pinkish pipes that rose partway up the canyon wall, toward what looked like an old mine building.

Water in motion is a powerful force. On the banks of a creek at the bottom of a hill near my house in Connecticut are the ruins of a nineteenth-century sawmill. The creek doesn't look like much, but a century and a half ago, at least, it moved fast enough to turn a small waterwheel, and the waterwheel turned a big saw blade, and the saw blade cut planks from chestnut trees that my town's early residents had felled in the hills above the creek. In the late 1800s, people began connecting waterwheels not just to saws and millstones but to electricity-generating turbines. A simple turbine produces electricity by moving a magnet along a wire, converting the mechanical energy of the moving magnet into an electric current in the wire. A simple turbine produces electricity by moving a conductor, such as a wire, inside a magnetic field, causing an electric current to flow within the wire and converting the mechanical energy of the moving conductor into electricity. (A simple electric motor does the opposite, by using a current in a wire to move a magnet, thereby converting electricity into mechanical energy.) The first such power plant in the United States was at the edge of Niagara Falls—a prime location, because the kinetic energy in moving water is proportional to its volume and its speed and the height from which it drops, and Niagara Falls has lots of all those things. Because hydropower doesn't produce thick clouds of black smoke, it was sometimes known in the olden days as "white coal." It remains, by far, America's largest renewable-energy source that doesn't involve setting things on fire—although it carries other environmental costs, because it always involves interfering with water's natural flow.

The big pale-pinkish pipes I saw rising above Shoshone were the power plant's penstocks—its man-made Niagara Falls, through which a steady stream of water falls toward the plant's turbine. I assumed that the penstocks must be fed by a stream or a lake on top of the canyon, but I was wrong about that; they're connected back to the river itself, through a tunnel whose intake is two and a half miles upstream, at a dam that's also next to the highway, just above Shoshone Falls. The tunnel was bored through the canyon walls between 1907 and 1909, and there's enough of a drop between its inlet and its outlet to generate fifteen megawatts. That electricity is used by customers in the service area of the power company that owns the plant, which has the right to divert 1,250 cubic feet of river water per second—more than 900,000 acre-feet per year. The plant has a priority date of 1902, making it the oldest large water right on the river. It's entitled to take its entire allotment before any junior holder, including metropolitan Denver, gets any, and for that reason it has been called "the most powerful water right on the Colorado River." Denver Water's Dillon Reservoir system is junior to it by almost forty-five years.

Decreed water rights can be lost through abandonment, so when

Shoshone suffered a mechanical failure a few years ago its owners rushed to make repairs. The plant is especially beloved on the West Slope because its "return flow" is close to a hundred percent, meaning that, even though the plant has the theoretical power to call out the largest city on the East Slope, essentially all the water it uses goes right back into the river, after passing through the plant's turbines. When the river is low, Shoshone sometimes empties the streambed between its intake and its outlet, but because the water isn't consumed downstream users are unaffected.

Shoshone's priority date has always made it seem a little bit like a tiny country with a large nuclear arsenal, although in 2013, after two very dry winters, Denver Water and the energy company that owns the plant negotiated a deal that relaxes some of the requirements of the prior-appropriation rules in times when water is short. That deal is too complicated to describe, but it demonstrates a larger truth about the Law of the River, which is that in periods of true stress even century-old legal precedents are not necessarily immutable—a good thing to keep in mind when people who live west of the Continental Divide talk about cutting off water to people who live east of the Continental Divide, or when people in Los Angeles talk about allowing Phoenix and Tucson to disappear. In a genuine water crisis, a small power plant in a canyon in a sparsely populated area would not be allowed to make the state capital uninhabitable.

WHEN I WAS SIXTEEN, I spent two weeks backpacking and rock climbing in the San Juan Mountains in southwestern Colorado, in what later became the Weminuche Wilderness area. The trip was organized by a now defunct outfit that patterned itself on Outward Bound. The following summer, a high school classmate and I went back by ourselves, after somehow convincing our parents that two not entirely trustworthy teenagers ought to be allowed to spend a fortnight in the wild without supervision. My mother told me, many years later, that she didn't know why they'd let us go, with our backpacks and climbing rope and freeze-dried mashed potatoes and no tent; we'd just somehow seemed too determined to be resisted. I'd shown my father on a topographical map more or less where we meant to go, but that was the extent of our planning. There were no cell phones or GPS devices or Park Service websites, and we hadn't read a guidebook about the region, and there was no one to check in with before we set out. Today, on Google Earth, I can zoom down on jade-colored mountain lakes that must be ones we camped beside, but at the time I felt almost as though we'd crossed into another dimension and that we'd become invisible to the rest of the universe. We explored canyons, and traversed snowfields, and camped where we wanted to, and climbed Storm King Peak. Our trip began and ended on a coal-burning narrow-gauge train that runs between Durango and Silverton, an old mining town. The train let us off—and picked us up again two weeks later—at Elk Park, a whistle-stop on the banks of the Animas River.

The Animas is a tributary of the San Juan, which is a tributary of the Colorado. It made headlines during the summer of 2015 when a contractor working for the U.S. Environmental Protection Agency, while attempting to prevent tainted water in an abandoned gold mine near Silverton from contaminating a nearby stream, accidentally broke through a barrier that had been semi-successfully holding the tainted water back. Several million gallons containing a variety of heavy metals and other contaminants spilled into the Animas and temporarily turned it bright orange. The plume eventually reached the Colorado, at the point where the San Juan flows into Lake Powell.

During the sluice-box era, gold miners had to cope with a scarcity of water; later, when mining moved underground, they had to cope with a surplus. Subsurface mines of all kinds almost always penetrate water tables, and while the mines are in operation groundwater seeps into them and has to be pumped out—a process known as "dewatering." In fact, the world's first commercial steam engines were used, in early-eighteenth-century Britain, not to power locomotives but to dewater coal mines—an innovation without which the Industrial Revolution as we know it could not have occurred. When subsurface mines of all sizes are abandoned, as they inevitably are, groundwater seeps in unimpeded and eventually fills all the empty spaces below the elevation of the water table. Various chemical processes then cause bad things to leach out of the newly exposed rocks, and a number of other factors—some natural, some not—make everything worse. The result is a phenomenon known as "acid mine drainage," which can contaminate both groundwater and surface streams. That's what the EPA was trying to deal with when the Animas accident occurred. That spill was relatively minor and short-lived, as environmental catastrophes go, but it ominously foreshadowed a far larger problem: Colorado alone contains more than twenty thousand abandoned subsurface metal mines, and many of them pose at least a potential threat to the Colorado River or other freshwater sources—in some cases, on a vastly larger scale.

I feel fortunate to have grown up during an era when the general level of adult anxiety about the safety of children was low. On backpacking trips organized by the summer camp I attended in Colorado in the 1960s, my fellow campers and I explored abandoned mine buildings, ate lunch on abandoned heaps of yellowish mine tailings, pilfered souvenirs from abandoned nineteenth-century miners' cabins, and even tiptoed short distances into abandoned mine tunnels—all while our counselors, who seemed ancient but must have been roughly college age, looked on without concern. Young people no longer routinely have access to adventures of that type. (The camp I attended is still flourishing, but I see from its website that when campers there take horse trips now they wear bicycle helmets, not cowboy hats.) The main thing I remember about those derelict mines, other than the joy of playing in them, is how often we came across them, almost everywhere we hiked. And, for the most part, they're still there, posing an unquantifiable but potentially enormous environmental threat, especially to water.

As mining technology has advanced—that is to say, as the efficiency of mining has increased—a number of its environmental impacts have grown more dire. When I followed Highway 24 up the Eagle River on my way to Aspen and Redstone, I came within a few miles of the Climax mine, the world's largest source of molybdenum. (Molybdenum is a metal with a very high melting point; it's used primarily in making high-strength steel and other alloys.) Climax opened in 1915, closed in 1995, and reopened, as an open-pit mine, in 2012. If you look down on the complex from the air, what you mainly notice, other than the vast scar of the mine itself, are the enormous "tailings ponds," which were created to hold wastes produced by Climax and other mining operations in the area. Open-pit mines require dewatering on a huge scale. The pumps often lower the local water table by hundreds of feet during the years when the mines are in operation, and the extracted water becomes a waste product that has to be disposed of, along with the waste products of ore-processing itself. (Gold producers go through a lot of cyanide.) Then, when the pumping stops, the drained cavities refill, creating the potential for toxic spills far larger than the one that contaminated the Animas. Some of the tailings ponds near the Climax mine have been partly reclaimed or remediated, in one case by using "biosolids" produced by the wastewater treatment facilities of Steamboat Springs and other municipalities. All are situated near the headwaters of important surface streams, including tributaries of the Colorado.

It's impossible not to be appalled by the damage that subsurface mining does to landscapes, ecosystems, rivers, aquifers, and human lives—although one's sense of outrage must necessarily be tempered by one's knowledge that the comforts of modern life have been fabricated pretty much entirely from the products of such mines. A cell phone alone can contain aluminum, antimony, beryllium, bismuth, bromine, cadmium, copper, gold, iron, lead, mercury, nickel, palladium, silver, tantalum, zinc, and many other extracted substances, and most of the energy that the phone and its mobile network depend on comes from the ground as well—and every year or two we get rid of our still-functional phone and buy a new one. Many of the technological wonders that we think of as solutions to our gathering environmental problems actually exacerbate other environmental problems—many so-called green technologies also depend on the extraction of a long list of mined exotic substances. In an op-ed piece in *The New York Times* in 2015, David S. Abraham—whose book *The Elements of Power*, published the same year, explored our growing dependence on rare metals—wrote, "They are embedded in the tallest wind turbines, the smallest computer chip and in every battery. If we succeed in meeting the International Energy Agency's minimum recommendation of nearly quadrupling wind power capacity by 2030, growing electric vehicle use more than 100-fold by 2025, and increasing battery efficiency, specialty metals like dysprosium and cobalt will take center stage." One consequence of our growing appetite for technological wonders and our unease about the environmental and human costs of producing them has been a shift in the worst of the direct damage to places we can't see—to parts of the world where labor and land are cheap, and where regulatory oversight is minimal. Electric and hybrid-electric cars (and cell phones and laptops and tablets and fitness trackers and untold numbers of other gadgets) depend on lithium-ion batteries, which depend on lithium, which increasingly comes from a single vast deposit near a Bolivian city that, in 2010, the country's president described to my *New Yorker* colleague Lawrence Wright as "a symbol of plunder, of exploitation, of humiliation."

. . .

FOR SOME TIME, the most notable form of subsurface mining in the
Colorado River Compact states has been the extraction of oil and natural gas by the method known as "hydraulic fracturing" or "fracking."
Garfield County, in western Colorado, contains thousands of fracked
natural gas wells, many of them within a mile or two of the river. You
can easily see them from the air: they look like homebuilding lots on
which the site work is nearly finished but the homes have yet to be
built, and there are so many of them that entire sections of what used
to be wild country look like nascent residential subdivisions.

In fracking, a shaft is drilled into a deep hydrocarbon-bearing rock
formation, and then water containing various additives is injected at
high pressure to create cracks in the rock, freeing trapped gas. The additives include chemicals that are used to control things like the viscos-
ity of the fracking fluid, along with sand or something similar, which
lodges in the cracks and keeps them propped open, like a foot in a door.
Some of the added chemicals are nasty, although the amounts are tiny:
fracking fluid is almost entirely water. The actual drilling and injection
doesn't last very long—usually a matter of days or weeks, for a well that
might produce gas for twenty-five or thirty years. Modern fracking
technology is so effective and efficient that it has transformed global
energy markets by creating gas and oil gluts that have pushed down
prices by mind-boggling amounts. (In the United States, a gallon of
gasoline cost less in 2016, in real dollars, than it did in 1985; in fact,
the cost of driving a mile in a gasoline-powered car isn't all that much
higher today than it was when I got my driver's license, in 1971.) Government estimates of the quantity of recoverable oil and gas remaining
in the world are actually significantly higher today than they were
twenty years ago, when one of the main topics of anxious debate was
when we would run out.

Once a gas well has been fracked, water stops going in and begins coming out. Some of this is returning fracking fluid, called "flowback," and some is groundwater that was formerly trapped in the same deep formation as the hydrocarbons, called "produced water." In some parts of the country, produced water is potable, or close to it; in others, it has to be treated and recycled or disposed of, sometimes by pumping it back into the ground. There are places in the United States where oil and gas production has caused the contamination of aquifers that people depend on for drinking water; in Garfield County, the gas-bearing formation is separated from shallow groundwater by a thick impermeable layer—the same one that kept the trapped gas from escaping many millions of years ago—and it's so deep, in some cases almost two miles down, that tapping it for water alone would never have been economical. Well sites in Garfield County are so densely spaced that the larger drilling companies have been able to build pipe networks to aggregate produced water from multiple wells for treatment, disposal, or reuse (sometimes in fracking fluid for new wells). And some cities in the county enthusiastically sell freshwater to drillers, partly in order to maintain the throughput volumes they need to keep their expensive municipal water-treatment facilities functioning. Since 2009, produced water has been governed by the priority system. Overall, the oil and gas industry's water consumption is small, even by comparison with municipalities. All industrial uses in Colorado add up to less than one percent of the state's total water use, and oil and gas account for a fraction of that.

On our flight to the headwaters, Jennifer Pitt, David Kunkel, and I flew directly over Boulder's principal source of electricity, Valmont Generating Station, which mainly burns coal. (Coal supplies more than eighty percent of Colorado's electricity; from the air, we looked down on Valmont's stockpile, which covers five acres and is maintained by freight trains that make the circuit from mines on the western side of the state—some of which I saw on my drive west.) If Valmont ever converted entirely to natural gas, most Boulder residents would view the change as a gain for the environment, because burning gas adds significantly less carbon to the atmosphere than burning coal does. But switching from coal to gas means you have to have gas—and to get it you have to get it out of the ground.

Natural gas is sometimes referred to as a "bridge" fuel to wind and solar—a methadone-like aid that can help to ease us away from our addiction to coal and oil—and its reputation even among environmentalists has lately been so high that at a conference in Washington, D.C., a few years ago I heard a speaker refer to it as "renewable." But natural gas is still a fossil fuel, and the fact that it costs very little right now has been as hard on the wind and solar industries as it has been on gas producers. The question typically posed by people concerned about climate change is whether we have the will, as a nation, to make a sufficiently large investment in (actual) renewable energy sources. But the real question is whether we have the will, as a species, to leave a sufficiently large fraction of the earth's abundant remaining fossil fuels in the ground forever, entirely untouched. That's a different question, and it's one that, in various ways, we all answer no to every day.

FRACKING HAS BEEN in the news so often that most people assume it's a recent invention, but the basic technique has been in use for more than half a century, and the practice of stimulating natural-gas production by breaking up subterranean rock formations is a century older than that. What has changed is the efficiency of the methodology. (People tend to think of technological innovation as a force only for good, but environmental problems innovate, too, and usually they have better funding.) Oil and gas companies employing current equipment and techniques are sometimes able to drill dozens of wells from a single pad, and they can tap widely separated regions in mineral-bearing formations without beginning directly above them—basically, by drilling sideways.

An early attempt to improve on fracking's efficiency was made in 1969, in a mountain valley a short distance from the Colorado River. It was part of Plowshare, a federal program whose purpose was to develop non-military uses for nuclear weapons—with which, then as now, the United States was ominously oversupplied. Plowshare was established in 1958, and when the government announced its creation the Atomic Energy Commission generated or received many proposals for making peaceful use of atom bombs: increasing subterranean water storage in Arizona and California; replacing the Panama Canal; excavating a harbor for large ships in northern Alaska; digging aqueducts to connect inland rivers in several states; dredging a channel through a barrier island off the coast of North Carolina; stimulating geothermal-energy production in a national forest in New Mexico; deepening the Bering Strait; mining coal, copper, and gold; building dams; building a second canal across Cape Cod; building a canal across Israel to connect the Mediterranean Sea and the Gulf of Aqaba; storing fuel oil under the seafloor off the coast of Guam; solving a frustrating sewage-disposal problem at Lake Tahoe; blowing up mountains that complicated the construction of highways and railroads in the western United States; and recovering petroleum from the Athabasca bitumen formation in Canada (better known as the Alberta tar sands). The AEC—along with the Bureau of Mines and the Nuclear Cratering Group of the U.S. Army Corps of Engineers—also believed that atom bombs might constitute an economical replacement for then conventional fracking techniques in the extraction of natural gas.

The 1969 test of that idea was called Project Rulison. It took place six miles from the town of Parachute, which straddles the Colorado River a little over forty miles downstream from Glenwood Springs. What was then known as the Los Alamos Scientific Laboratory (operated at the time by the University of California) made an unclassified seven-and-a-half-minute newsreel-style public-information film about the test before it occurred. The film's narrator describes Project Rulison as "a unique experiment" to be conducted by "a joint industry-and-government team," and explains how it's going to work: "A nuclear explosive, equal to forty thousand tons of TNT, will be used to shake loose a great natural-gas reserve locked tightly in a formation called the Mesa Verde. . . . Its purpose is to test the feasibility of using nuclear stimulation techniques to develop commercially a natural-gas-bearing field in a low-permeability formation." One of the government's industrial partners in the experiment was the Austral Oil Company of Houston, which owned the mineral rights to more than sixty thousand acres. The explosive device was roughly two and a half times as powerful as the bomb the United States dropped on Hiroshima.

The film is available on YouTube, and it's worth watching, not least because it will remind you of the kinds of cars that people drove in 1969. "At stake is an estimated 110 billion cubic feet of gas per section in place at the experiment site," the narrator continues. "If the Rulison test shows this gas can be extracted at a profit, it will open the way to recovery of vast gas resources not now accessible—eight trillion cubic feet in the Mesa Verde formation alone." An "exploratory test well" was drilled near "surface ground zero," partly to generate a baseline measurement of pre-blast gas production and partly "to determine the nature of the underground water table, if any." Scientists concluded that "there is no flow of water, or supply, that could be harmfully disturbed by the detonation." The actual blast shaft was drilled in a high valley above the river and was eighty-four hundred feet deep.

"The energy released by the nuclear explosion will melt and vaporize nearby rock and will fracture the rock beyond to a diameter of about 740 feet," the narrator explains. He continues, "As the cavity cools, the vaporized and melted rock will collect in a puddle at the bottom, and most

of the radioactivity will be entrapped here, as it solidifies." Gas was expected to accumulate in the cavity and in a rock-filled "chimney" that
formed above it; drawing it off would be as simple as sucking soda
through a straw. Producing the same effect with nitroglycerine, a fracking staple at the time, would require "nineteen million quarts" and entail
"nerve-shattering" safety problems, the narrator says. The government
already had a pretty good idea of how the test would turn out, because it
had detonated a similar nuclear device in a gas-bearing formation in
New Mexico in 1967, in a Plowshare project named Gasbuggy.

Project Rulison attracted protests, but not as many as you would
think. The detonation took place on September 10, 1969—three weeks
after Woodstock. On the *CBS Evening News* that night, Terry Drinkwater reported that "not everyone was happy" about the test, and that
roughly a hundred protesters had "marched on the observation tent"
(carrying signs that said, among other things, "Kill Nature for Gas?").
The demonstration was orderly, however, and the blast went off as
planned. "The earth shook like jelly, there was a muffled sound, blocks
and dirt shook loose from surrounding mesas," Drinkwater continued.
"In Grand Valley, a few bricks fell from a few buildings." Workers with
Geiger counters checked the site for contamination—"They found
none"—and the crew at the control center celebrated with champagne.

Prematurely, it turned out. The explosion freed what was estimated
to be 455 million cubic feet of gas. That was disappointingly less than
the experts had predicted, and, worse, the gas was rendered unusable by
high levels of tritium, a radioactive isotope of hydrogen.

THE RULISON SITE IS MARKED by a small concrete pad and an embedded plaque, which explains what took place there and warns passersby
that removing subsurface materials from below a depth of six thousand
feet is forbidden. I decided to try to find the plaque. I began with break-

fast at Shooters Grill, in Rifle, a river town roughly twenty miles up-stream from Parachute. Shooters is famous for encouraging customers to bring their guns with them and for employing waitresses who carry pistols. No one on the premises was armed when I was there—a disap-pointment—but my breakfast, an everything-you-can-think-of concoc-tion called a Six Shooter Skillet, was terrific. Thus fortified, I entered the latitude and longitude of the test site (available online) into the nav-igation app on my phone and drove into the hills.

My route took me past Battlement Mesa, a semi-unsuccessful-looking real estate development overlooking the Colorado—where "the American Dream is alive and well," according to its website—and a golf course whose entrance was marked by a small artificial waterfall. It also took me past many fracking sites, only a few of which were visible from the road. A sign at the edge of a broad empty meadow covered with brownish native grasses advertised "Ranchettes for Sale." There were also signs telling the drivers of gas-industry vehicles what to do and signs telling them what not to do. I passed a guy in an orange cap standing next to a car, and I passed some people having a picnic next to a stream, but I didn't see anyone else. The dirt road I was following was narrow, and it climbed steadily higher into an aspen-filled mountain valley, which was flanked, on the left, by an enormous rock face. The farther I drove, the more adamant the "No Trespassing" signs seemed to become—probably because the bullets that people had used to shoot holes in them seemed to steadily increase in caliber. Then, just a few hundred yards from the marker, my GPS-directed route was blocked by a locked gate. I knew from Google Earth that I could reach ground zero by ignoring my phone and continuing straight ahead, through another gate and past another bullet-riddled "No Trespassing" sign. But this last sign seemed to have been shot up by a howitzer, and the road beyond the gate looked more like a driveway than a road, and as I thought back to the guy in the orange cap I decided that he had looked at me, as I passed him, with something like suspicion. I chickened out and turned back toward the river.

Nineteen sixty-nine wasn't that long ago. How could responsible adults so late in the twentieth century have thought that using nuclear weapons to create natural-gas wells could possibly be a good idea? Freeing all the gas that was believed to be trapped in the Mesa Verde formation, assuming the Rulison results were representative, would have required more than fifteen thousand detonations of the same size. And yet three and a half years later the AEC tried again, thirty miles northwest of Rulison, on the other side of the river, at Rio Blanco. This time the principal industrial partner was Continental Oil Company, and there were three nuclear devices, totaling one hundred kilotons. They were placed in a single shaft, separated from one another vertically by roughly four hundred feet, and detonated almost simultaneously. Once again, the liberated gas was radioactive and there was much less of it than the experts had predicted. ("And such small portions!") If the AEC's scientists had believed that increasing the power of a nuclear blast by 150 percent would eliminate the resulting radiation, they were disappointed.

Rio Blanco attracted even less public outcry than Rulison had, probably because the country at the time was preoccupied by Richard Nixon: the Senate Watergate Committee began its hearings, which were televised nationally, the same day. A fourth natural-gas test, Wagon Wheel, was planned for a site in southwestern Wyoming but was never conducted. (It would have employed five hundred-kiloton devices.) Plowshare itself ended for good in 1975, and it's a useful reminder that big ideas often have a short half-life. The Gasbuggy, Rulison, and Rio Blanco sites are still monitored for surface contamination by the Department of Energy's Office of Legacy Management. According to that office, no radioactivity associated with the tests has ever been found in wells, aquifers, or other water sources in the surrounding areas, including the Colorado River.

GRAND VALLEY

A few miles downstream from Rifle and Parachute, De Beque Canyon opens into the Grand Valley, a broad basin a dozen miles wide and three dozen miles long. It's covered with closely spaced irrigated fields, which from the air make the valley floor look like an enormous green patchwork sock. The major population center is Grand Junction, so named because near the center of town the Colorado River (née Grand) is joined by the Gunnison River, a major tributary, which arises in mountains roughly 160 miles to the east-southeast. Farming in the Grand Valley began in the late 1800s, once white settlers had removed the Indians, and was made possible by irrigation. (The region receives only about eight inches of rain a year.) Early farmers transported water in barrels; the first irrigation canals and ditches weren't built until the 1880s. Construction of the modern irrigation system began a couple of decades later, based on a plan approved by James R. Garfield, a son of the former president, who was Teddy Roosevelt's secretary of the interior. The region was attractive to farmers, despite the lack of water, because the soil was good and the growing season was fairly long.

Irrigation water is drawn from the Colorado by the Grand Valley Diversion Dam, which was completed in 1916. The dam is a few miles east of the eastern end of the valley, and I drove right past it as I emerged from De Beque Canyon. It's a "roller gate weir," so named because the flow of the water is controlled by raising and lowering six enormous steel cylinders, or rollers—a technology patented by a German company shortly after the turn of the twentieth century. The dam is 550 feet from end to end, and each roller is 70 feet long and 7 feet in diameter. (A seventh roller, somewhat shorter, controls a sluiceway at one end of the dam.) The rollers are mounted between concrete towers with orange tile roofs, and they can be raised or lowered as river conditions change. "When raised, the openings allow for the passage of large objects such as trees and ice floes over the crest of the dam," a Bureau of Reclamation history explained in 1994. "The rollers may be used in any combination to maintain the proper water level regardless of the rate of flow of the river." When lowered, the rollers seal tightly against a sort of concrete trough, and the river runs over the top. The dam presents an unusual drowning hazard under certain conditions because, on the downstream side, water flowing over the rollers rebounds off the bottom of the channel and creates a swirling backward vortex, which doesn't necessarily look dangerous from the surface but can be impossible to swim out of and difficult even to recover bodies from. For that reason, roller dams are sometimes called "killer dams" or "drowning machines." There aren't very many of them, although other types of "low-head" dams—which also allow water to flow over them and pose similar dangers for similar reasons—are common all over the United States.

River water diverted by the Grand Valley Diversion Dam is channeled through a network of canals, reservoirs, and conduits, and serves roughly forty thousand acres of agricultural land. I drove across the valley toward the western limit of the irrigated area and stopped at the intersection of 10½ Road and Q¾ Road, not far from the corner of 11⁸⁄₁₀ and P. (Grand Junction's first north-south roads were named for their distance in miles from the Utah border, and the east-west roads were named with letters of the alphabet, and many of the gaps have been filled in with fractions and decimals.) I parked next to an alfalfa field that had recently been flood-irrigated. Flood irrigation consists of watering a crop by inundating it, and it's probably the oldest form of large-scale irrigation. A dozen feet from my car was a big supply pipe, which ran along the ground at the upper end of the alfalfa field. To irrigate his crop, the farmer had opened valves in the pipe and kept them open until water had flowed all the way to the field's far side, which was slightly lower in elevation. Flood irrigation is not efficient, since as little as half of the applied water ends up in the crop. (Much of the rest runs off or evaporates.) The road's shoulder, between the pipe and my car, was several inches deep in water that had leaked from the pipe or flowed the wrong direction, leaving some very well-irrigated gravel.

I drove back to the eastern end of the valley and bought some potentially addictive blood plums at a farm stand in Palisade, then visited Mesa Park Vineyards, on C Road between 33 and 33½. Mesa Park is owned by Brooke and Brad Webb, who have a young daughter, and by Brooke's parents, who live with them. The Webbs both worked in finance in Denver, and they both liked wine, and in 2009, when they noticed a real estate ad for an eight-acre vineyard on the west side of the state, they decided to change their lives. Brooke's father is a retired aerospace engineer; during his career, at Lockheed, he acquired skills that have turned out to have applications in wine-making. I parked behind the Webbs' house, next to a bright red barn with white trim. The barn contains their fermenting room and wine-making equipment, as well as a wine-bar-like tasting room. Brad was chatting with customers in the tasting room, so Brooke and I went behind the barn and sat in the shade in some old white plastic chairs.

"We're open seven days a week, all year," she said. The Webbs grow only red grapes, so in order to sell white wine they have to buy grapes from other growers—and during the year of my visit they had had to buy their red grapes, too, because cold weather the previous December had killed almost all their grapevines. The growing season in the Grand Valley is two months longer than the growing season in the wine regions of New York State, making it almost the equal of California, but the winters can be tough. We walked along a strip of grass between trellises. "We're kind of at the edge of survivability for vineyards," she said. The vines on either side of the path were all new and wouldn't produce grapes until the following year. The vineyard's previous owners had had fifteen or twenty consecutive years of good crops, but the string of good harvests ended almost as soon as the Webbs took possession. In several places I could see remnants of the old vines, which were as thick and twisted and gnarly as the trunks of small, wizened trees.

We got into my car and drove around to the back of the property, so that Brooke could show me the canal that feeds their irrigation ditches. "The history of Palisade is all about water," she said. "A hundred percent of the water we use to irrigate comes from the Colorado River, and it would just be a desert here without it. We have fantastic pre-1922 water rights, and we can access the canal whenever we want to between May and November." (The Webbs are entitled to draw water from the canal; the right to divert water from the Colorado belongs to their water district, which operates the canal.) I parked on a dirt access road that runs between the Webbs' property and a neighbor's, and we walked along a gravel path next to the canal.

The canal was ten or fifteen feet wide, and the water in it was brown. On the near bank, on either side of a little wooden bridge, were several headgates. "This one's ours," Brooke said. "Number eighty-eight." Each headgate looked like a submerged miniature guillotine, the top of which protruded a couple of feet above the surface of the water. Below the surface was a steel plate that covered the opening of a submerged pipe leading to a farmer's irrigation system. The plate could be raised or lowered by turning what looked like a huge faucet handle, or a miniature steering wheel, at the top of the guillotine. When the Webbs are irrigating, they turn the handle on number 88, and water from the canal flows into the pipe, under the path, and into a cylindrical distribution box made of precast concrete sections. From there it flows into moveable pipes that the Webbs attach to the distribution box, and the pipes direct water into shallow ditches crisscrossing their vineyard. The canal is maintained by the irrigation district, and the water in it comes straight from the river. The Webbs' drinking water has a different source, Grand Junction's main municipal water utility, which draws most of its supply from reservoirs fed by smaller streams, plus a relatively small amount from the Colorado. Before distributing that water, the utility treats it for the usual contaminants and impurities.

We walked back along the access road, past a neighbor's pear orchard. Pears don't ripen on the vine, the way grapes and peaches do, and the main market for them is the baby-food industry. People who make hard cider, a popular product in recent years, are willing to pay more for pears than baby-food companies do, but many growers are bound by long-term contracts. I asked Brooke whether she and her husband had ever thought about switching to peaches—the crop that established the Grand Valley as an agricultural area. "No," she said. "I'm totally a wine enthusiast. But I guess what we've realized is that it's tough to be a farmer, and it's expensive, and the money isn't in growing grapes." Their vineyard isn't a loss leader, exactly, and it's more than just a showpiece, because the grapes they grow there go into their premium "estate" wines. But the foundation of their business is the tasting room, and even in good growing years most of their wine-making is done with grapes grown by someone else. The same is true for most of the twenty-odd other vineyards in the valley, and of the valley itself.

Wine-making is more nearly a part of the regional tourist industry, since one of its main economic contributions is helping to fill hotels and restaurants with people from other places. By doing that, it helps support other local businesses, including rafting companies and bicycle shops. (Grand Junction is a major mountain-biking hub.) One of the biggest tourist draws is the Colorado Mountain Winefest, which attracts thousands of visitors and has been held in Palisade each September since 1991.

The economic precariousness of the Webbs' new enterprise has increased their interest in water conservation. "When we moved in, the water wasn't being managed well," Brooke said. "Our neighbors said, 'We have a water problem, and you guys are it.'" They began digging new furrows between their trellises every year, and they cleaned out their ditches and lined them with plastic, to keep them from leaking, and they eliminated a wet area at the bottom of their vineyard, where excess irrigation water used to pool. "There are still issues with neighbors, although we now actually handle wastewater for several of them," she said. "A lot of the old-school farmers have a use-it-or-lose-it mentality and are not interested in the conservation-minded practices that we employ—although we're still flood irrigators. Drip irrigation would be more efficient, and there are a lot of grants available for that, so that's a goal we have. But we still estimate that we now use only about seven percent of our allotment."

IN 2013, the nonprofit organization American Rivers placed the Colorado at the top of its annual list of the ten most endangered rivers in the United States. The names on that list almost always change completely from one year to the next—the ranking has more to do with promoting general awareness than with tracking specific threats—although number two in 2014 was the Upper Colorado, a rare back-to-back repetition, and number one in 2015 was the Colorado again, this time the part of it that flows through the Grand Canyon. After the 2014 list came out, I spoke with Sinjin Eberle, who works in the organization's communications office. He's based in Colorado, which began working on a statewide water plan in 2013. "We have a million laws and regulations," he told me, "but we don't have an overarching plan for how water is managed throughout the state." He said that American Rivers and other environmental groups were working to make sure that several key principles were included in the final version, and that among the most important of those was an emphasis on conservation and efficiency, both in municipalities and on farms.

That seems like an obvious strategy, and it's easy to understand—if we're using too much water, we need to use less—but the issue is actually complicated, and the consequences can be counterintuitive. Bradley Udall, a senior water-and-climate-change research scientist at Colorado State University's Colorado Water Institute, told me, "Water is a super-confusing topic for many people, because the language and mental models we use around the efficient use of resources are almost always wrong in the context of water." Udall is a member of a family with a long, celebrated history of public service, including involvement in environmental issues. His father, Morris, was a congressman from Arizona for thirty years and a candidate for the Democratic nomination for president in 1976; his uncle Stewart was the secretary of the interior during most of the 1960s; his brother Mark was a U.S. senator until 2015; and his cousin Tom is a U.S. senator now. "My mother took me down the Grand Canyon when I was fifteen years old, in 1972," Bradley told me, "and it was a life-changing experience for me. I wanted to be a Grand Canyon river guide, and I later became one, for two years." He's been thinking about water ever since.

Water uses, Udall said, can be divided broadly into two categories:

consumptive and non-consumptive. When a Grand Valley farmer flood-irrigates a vineyard with Colorado River water, some of the water goes into the grapevines. That water, along with any water that evaporates from the field, can't be used again until it has worked all the way through the hydrologic cycle, usually by emerging as precipitation somewhere else, and for that reason it's referred to as "consumed." But not all of the farmer's irrigation water goes into the grapes or evaporates. Some of it runs off the end of the field and is channeled back into the ditch system, from which it can be diverted again, by other farmers—and that excess is referred to as "non-consumed." And some non-consumed water makes it all the way back to the river, earning a "return-flow credit" for the irrigation district. (Kent Holsinger told me that, on average, river water in Colorado is used more than half a dozen times before it leaves the state.)

In many cases, irrigation water that soaks into the ground can be considered non-consumed as well, because it helps maintain the local water table and replenishes aquifers that feed local wells and surface streams. Udall said, "Colorado high-mountain ranchers say that over-watering in the spring and summer isn't wasteful because it leads to delayed subsurface return flows, through groundwater, that keep our rivers higher in the fall and winter than they would be otherwise. And in many cases they are right." Requiring such ranchers to adopt more efficient irrigation techniques can have the perverse effect of increasing the proportion of water they consume, by enabling them to irrigate additional acres with water that used to be non-consumed.

"Efforts to improve water efficiency in agriculture almost always lead to increases in the consumed fraction," Udall continued. "On an individual field, they make it look like we are using water better, but they actually move us in exactly the wrong direction." The authors of *Managing California's Water*, a report published in 2011 by the Public

Policy Institute of California, wrote: "Irrigation improvements can actually *increase* net water use by crops, by allowing either more intensive use of irrigation water on a given field (which raises both yields per acre and net water use per acre) or more extensive use of 'saved' water on nearby fields that were previously less irrigated." Irrigation improvements can harm entire ecosystems, far beyond the cultivated area. Michael Cohen, of the Pacific Institute, told me, "In a lot of places, the environment has come to depend on slop in the system." Water that flows away from irrigated farmland sometimes sustains neighboring wetlands, and if farmers become more efficient the wetlands disappear. Increasing efficiency also does nothing to address over-allocation—just as becoming a smart shopper won't get you out of debt if your total spending stays the same. Indeed, increasing efficiency can make over-allocation more dangerous, by allowing the total number of users and uses to grow, thereby eventually promoting additional consumption, and by making it harder for all users to cut back.

Nevertheless, Cohen told me that he finds dismissals of the value of efficiency and conservation to be "an interesting theoretical construct that has limited contact with the facts on the ground." He continued, "Typically, irrigators implement efficiency improvements because somebody is paying them to do so, so that the investor—such as the Metropolitan Water District of Southern California or the San Diego County Water Authority—can reap some or all of the conserved water. It's a zero-sum game only from the very narrow scope of the water being transferred. From a system-wide perspective, the efficiency gains and transfer of the conserved water mean that the recipient no longer looks to developing new supplies—say, through desalination or additional diversions—and improvements in efficiency and conservation mean less anthropogenic water consumption overall. And there are a host of corollary benefits in water quality and energy use. For example, as the residents of Las Vegas consume less water, the Southern Nevada Water Authority pumps and treats less water from Mead, and then has less wastewater to treat before it discharges it back into the lake."

Still, the long-term environmental consequences of efficiency improvements of all kinds depend on what happens to the savings. Running a kitchen faucet or flushing a toilet in a municipality with a modern sewage-treatment system is mostly non-consumptive, because the wastewater is treated and used again. Watering a lawn, in contrast, is almost entirely consumptive, because as far as the municipality is concerned that water disappears as soon as it hits the ground. Popular conservation schemes can sometimes merely substitute consumptive uses for non-consumptive ones. Imagine a municipality with so-called block, or tiered, water rates, which are kept low below some threshold, to make ordinary household use affordable, and then rise dramatically, to discourage people from casually doing things like washing cars and watering grass. If residents of that municipality now install bathroom and kitchen fixtures that use less water, they shrink their non-consumptive use—by reducing their wastewater return flows back into the system—while simultaneously making watering their lawn and washing their car more affordable, since now they can do it with cheaper water. They're using water more efficiently, because they're receiving more value from every gallon; but they've shrunk the available supply of the local water system. The same efficiency efforts can cause operational problems for municipal sewage-treatment facilities, which require volume and dilution in order to function properly. In 2015, prompted by the drought, Californians were remarkably successful at cutting domestic water use, but an unanticipated consequence in many cities was clogging, corrosion, intrusion by tree roots, and other damage within those cities' waste systems, which were not designed to function without big flows to keep everything moving. These are problems that can be overcome, but overcoming them costs money. Conservation increases a municipality's per-gallon cost of financing, building, and maintaining the infrastructure that moves water in both directions, and consumers inevitably complain if their bills go up as their consumption goes down.

Waste, paradoxically, is a kind of reservoir. If the residents of a suburb routinely water their lawns, they can stop during a drought. But once they've replaced their Bermuda grass with cacti and gravel, and once the water that formerly ran through their sprinklers has been redirected to bathrooms and kitchens in brand-new subdivisions, the system not only is spread across more water users but also is more vulnerable in dry periods, because it contains less slack: when you increase the human utility of a gallon of water, you also increase the human impact of losing that gallon. That's not an argument against using less water, but it's a reminder that issues related to the consumption of natural resources are seldom simple.

One of the many complexities of water law in Colorado is that people in one river basin who use water that originated in another river basin—such as residents of Boulder who wash their cars with water that was transported by tunnel from the other side of the Continental Divide—are entitled to use that water "to extinction": they have no legal obligation to return any unused portion to the originating river system. That sounds like a bad thing, but it has often been a good thing, because in times of plenty the unconsumed portion has typically flowed into other streams, or helped to recharge aquifers by soaking into the ground, or performed other useful functions. As populations have grown, however, municipalities have begun taking steps to become more efficient in their use of trans-basin water, despite having no legal obligation to do so. That sounds like a good thing, but it can often be a bad thing, because—as Jones and Cech write in *Colorado Water Law for Non-Lawyers*—"water reuse programs, . . . although efficient for the city involved, have the potential to substantially reduce the flows" of rivers that currently depend on the waste of municipal users. Reducing waste, in such cases, can create problems for downstream users—especially farms. Reusing in-basin water can actually be illegal, if the reuse causes mandated return flows to fall.

Udall told me that, for all such reasons, improvements in water efficiency behave differently from improvements in energy efficiency—but, actually, the two are similar, and they pose essentially the same environmental dilemma. In both cases, attempting to reduce total consumption by making current consumption more productive has impacts that can be the opposite of the intended ones, because—with energy as with water—we almost always reinvest our savings in additional consumption. One of the great difficulties in truly addressing climate change is that the problem is cumulative (because every bit of fossil fuel we burn makes what is essentially a permanent addition of carbon to the atmosphere), while the most popular solutions are not cumulative (because when we "save" fossil fuel by modifying wasteful behavior we inevitably use our savings to do something else). A few years ago, I made a serious effort to get better about turning off the lights in my house, and my wife's and my electricity consumption went down by a noticeable amount. But our overall energy consumption didn't fall, because the money we saved on our electric bills helped to pay for a big anniversary trip that we took to Europe, and that means that the real impact of our reduction in household electricity use was merely to transform natural gas into jet fuel. As we get better at doing things, we do more things.

A friend recently sent me an article from *Slate* about a superlightweight structural material being developed by Boeing. The article said that the material, "microlattice," is so light yet so strong that "its use could have some great economic and environmental benefits" by making airplanes "significantly more fuel efficient." But, as is also the case with water, economic benefits and environmental benefits usually pull in opposite directions. Making airplanes more efficient makes flying less expensive, so we do more of it—and we know that really happens because we've already run the experiment. Today's passenger jets are something like seventy-five percent more fuel efficient than the passenger jets of the early 1960s, yet the total amount of energy consumed by aviation today is vastly greater than it was then. Flying was an expensive luxury when it was less efficient; the more efficient it has become, the less it has cost and the more of it we have done. In 2016, I flew from New York to Ireland for twenty-five dollars less, in unadjusted dollars, than it cost me to make the same trip in 1993. When we use finite resources more efficiently, we make their consumption less costly to ourselves. The modern myth is that by maximizing economic value we can break the link between consumption and its consequences—that there's a sort of Moore's Law of human comfort. But at some point, inevitably, we will reach an absolute limit.

6.

SALT, DRY LOTS,
AND HOUSEBOATS

Near the western end of the irrigated portion of Grand Valley, I-70 diverges from the river, but I returned to the water, a few miles over the Utah line, by taking the exit for Route 128, which is also known as the Upper Colorado River Scenic Byway. It's one of the most beautiful highways I've ever driven on. Parts of the roadway were built on an old pack trail, and the whole thing is so sinuous and narrow that big trucks aren't allowed. It follows the river past squat mesas and a weathered wall of spindly pinnacles called the Fisher Towers, which look like enormous red stalagmites, then descends into a sandstone gorge that could be a preliminary, reduced-scale draft of the Grand Canyon. Route 128 ends in Moab, a town whose transient population swells during the summer and on decent weekends in all seasons. Moab is an assembly point for two national parks, Arches and Canyonlands, and in nice weather the cars and RVs on the main drag bristle with kayaks and mountain bikes.

About halfway between I-70 and Moab, I passed the remains of a historic wood-decked suspension bridge that crossed the Colorado until 2008, when a six-year-old boy accidentally burned it down while playing with matches. (All that's left of the bridge today are the steel cables that held it up.) About a mile and a half upstream from the bridge is the Colorado's confluence with the Dolores, a tributary that arises more than two hundred miles to the south, in San Juan National Forest in southwestern Colorado. The Dolores was mostly wild until the 1980s, when the Bureau of Reclamation built a dam roughly twenty-five miles downstream from its headwaters. The dam and its reservoir were named for the town of McPhee, Colorado, the remains of which now lie at the bottom of the reservoir. The town was created in the early 1900s to house the employees of a large lumber company, which operated there until the 1940s.

The Dolores has had a surprisingly large impact on western water use—and the reason is salt. The Colorado is saltier than most rivers in the United States, and most of its salt has the same source as most of the salt in the oceans. As rain falls through the atmosphere, carbon dioxide in the air dissolves in it, forming small amounts of carbonic acid. When this mildly acidic precipitation then flows over land, it leaches salts and minerals from rocks and soil, and eventually carries them into surface streams, which carry them to the sea. Hydrothermal vents in the ocean floor and undersea volcanic activity add more, and the total accumulation is enormous. According to an estimate cited by the U.S. Geological Survey, if you removed all the salt from all the world's seawater and spread it evenly on land, it would cover the entire non-ocean surface of the earth to a depth of more than five hundred feet.

The concentration of salt in the oceans has been roughly the same for hundreds of millions of years, and the reason is that, even though salt from sources on land continually flows into the sea, an approximately equal quantity continually settles out, mainly in the remains of dead organisms, which sift to the bottom and become part of the ocean floor. Surface streams, by contrast, vary greatly (and often fluctuate) in their salinity, mainly because they don't all flow over terrain with the same chemical composition. The Colorado is high in salts because large sections of its basin consist of relatively soluble sedimentary rocks, which were formed partly from salty stuff that accumulated at the bottom of primordial seas, and other sections consist of porous volcanic rocks, which contain salts and minerals that originated deep in the earth's crust. Agriculture, gardening, residential water softening, municipal wastewater treatment, highway deicing, and innumerable other human activities add still more salt to the river, and the concentration generally increases as the Colorado moves south, toward Mexico.

The Dolores used to be a major salt contributor. About 150 miles downstream from McPhee Dam, it flows across Paradox Valley—so named, by a nineteenth-century geologist, because the river crosses it from side to side, rather than end to end. (The Colorado itself does something similar in eastern Utah.) It does that because the river already existed when tectonic movements began slowly lifting the mountain ridges that form the sides of the valley, and the flowing water eroded a path through the ridges as they rose. Those same tectonic movements also created, very close to the surface, an enormous subterranean salt dome—the remnant of an ancient sea. Parts of the dome are more than two and a half miles thick, and as the Dolores flows across the valley it picks up salt from groundwater perched on top of it, as well as from the ground itself.

The salinity of the Colorado became an international issue in the early 1960s, when the river's salt load reached a level that killed the crops of irrigators in Mexico. I'll have more to say about that issue later, but one eventual element of its resolution was the construction, in 1991, of a salt-removal facility on the Dolores River. The facility, the Paradox Valley Unit, is operated by the Bureau of Reclamation and is situated just outside Bedrock, an unincorporated town whose general store appears briefly in the movie *Thelma and Louise*. "At the Paradox Valley facility," Stephen Elliott, a staff reporter for the *Telluride Daily Planet*, wrote in 2015, "extraction wells between 40 and 70 feet deep along the Dolores pull brine out of the groundwater beneath the river, process it and pump it three miles across the valley." There, the processed water is fed under high pressure into a deep injection well, which pushes the brine into a geological stratum that lies just under the bottom of the salt dome, two and a half miles down. The dome acts as an impermeable cap, preventing the injected brine from returning to the surface. Since 1991, the PVU has injected more than two and a half million tons.

The PVU has beneficially reduced the salt content of the lower Colorado; the Bureau of Reclamation has estimated that every ton of salt it removes prevents $173 worth of damage downstream. But it has also caused roughly six thousand earthquakes, the largest of which, in 2013, had a magnitude of 4.0. The connection between the earthquakes and the injection well is undisputed, since the bureau's scientists knew before the project began that deep injection would generate many small tremors, and since tremors of any kind hardly ever occurred in the area before 1991. It's also fairly clear that the seismic activity has increased in intensity over time, as the injection well, acting like a basketball pump, has steadily raised pressures under the dome. To date, the earthquakes have seldom been noticeable, and the bureau's scientists have said that they will never be much stronger than the biggest ones that have occurred already. But the facility's engineers, nevertheless, have decreased the pressure at which they inject brine into the formation, and the bureau is exploring alternatives, and at some point they will have to make further changes. The possibilities include drilling a replacement well in a new location, assuming a suitable one can be found; dealing with extracted brine by impounding it in evaporation ponds and then somehow disposing of the residue, an approach that would create environmental issues of its own and might be prohibitively expensive; and pushing the problem downstream. The alternatives probably don't include doing nothing—a familiar paradox of civilization.

. . .

BETWEEN MOAB AND GLEN CANYON DAM, in northern Arizona, the Colorado is only minimally accessible by car, because most of its course lies deep within enormous tracts of protected land. A month after my trip across Colorado and eastern Utah, I took the long way back to the river, by flying to Phoenix and driving north for four and a half hours. I stopped briefly in Flagstaff, at roughly the halfway point on my route. Flagstaff is economically dependent on the Colorado—it's the last convenient source of provisions for people traveling to the Grand Canyon from the south—but it gets no water from the river. It was established in the early 1880s, primarily as a watering stop on the Atlantic & Pacific Railroad, but even in those early years, when its population could be measured in dozens and then in hundreds, it struggled with access to water. Today, Flagstaff has a population of roughly seventy thousand, and it draws its drinking water from a varied and not entirely reliable assortment of reservoirs, wells, and storage tanks. In a government building downtown, I met with Liz Archuleta, a longtime member of the board of supervisors of Coconino County, who told me that colleagues of hers had made a preliminary study of the possibility of supplementing the area's water supply by building a pipeline from Lake Powell, 130 miles to the north, but had discovered that this seemingly simple idea was more complicated than anyone had anticipated. ("Ah, that would violate the Law of the River.") They hadn't given up, necessarily, but they also hadn't placed an order for conduit.

Coconino County is the second-largest county in the United States by area; it's also one of the driest. Many places beyond the reach of the city's water utility have no water at all, because surface streams are nonexistent or unreliable, and because what groundwater exists either is undrinkable or lies too deep for economical pumping. (Lifting water out of the ground takes energy. An acre-foot weighs almost fourteen hundred tons.) "The majority of the county's residents outside of Flag-staff itself haul water," Archuleta said. People who live in what are known as "dry lots" or "dry-lot subdivisions" have to order their water from a commercial supplier, or buy it at a pipe stand, often a long dis-tance away, and carry it home themselves in a big plastic tank that fits on a trailer or in the bed of a pickup truck. In Ash Fork—fifty miles west of Flagstaff, at the eastern edge of Yavapai County—water-filling stations routinely have long lines.

Even people who receive water from a local utility also sometimes buy hauled-in water to supplement what comes from their faucets or to circumvent watering restrictions, and some people who have drilled wells don't use them because the water they provide contains high lev-els of things like arsenic, which occurs naturally in deep formations. Archuleta said, "Population growth here is not dependent on water, in the sense that we cannot consider the availability of water when ap-proving or disapproving land-use applications. And you can split your property into as many lots as you like, without a subdivision plan, as long as the lots are at least two and a half acres. You don't have to prove that you have water or infrastructure or anything else. You just go and record the new lots at the recorder's offices." A number of years ago, the Arizona legislature made it illegal to develop dry lots, but a moratorium on new regulations, which was first imposed by the governor in 2009 and has remained in effect ever since, has prevented enforcement.

PART OF THE REASON American Rivers placed a section of the Colo-rado at the top of its list of the country's most endangered rivers in 2015 was a proposal by the Navajo Nation to build a resort and gondola tramway on tribal land in haul-in territory on the east rim of the Grand Canyon, near the river's confluence with the Little Colorado. That project has apparently stalled, although American Rivers cited other threats, too, among them a proposal by an Italian company to build a huge tourist and residential development a short distance downstream from the tramway site, on a parcel it has assembled near the otherwise uncelebrated town of Tusayan, Arizona. For the tribes and for residents of places like Tusayan, the river and the canyon are among the few potential sources of income in a remote, water-starved region that offers little else in the way of economic opportunity. Hopi and Navajo reservations cover virtually all of the northeastern corner of Arizona, beginning just beyond Flagstaff, and more than half of their residents haul water and have incomes below or near the poverty line.

The Colorado River Compact mentions Indians only once, in Article VII, which reads, in its entirety: "Nothing in this compact shall be construed as affecting the obligations of the United States of America to Indian tribes." (That sentence was added at the insistence of Herbert Hoover.) The tribes, collectively, could conceivably make priority claims to all the freshwater in America, since Indians were using streams long before the invention of the sluice box. That isn't the way things have worked out, quite obviously, but the tribes' water entitlements are still potentially huge. In 1908, the Supreme Court, in *Winters v. United States*, decided, in effect, that the government wouldn't have created Indian reservations if it hadn't meant for Indians to have access to enough water to support the crops on which the government now expected them to survive—a legal concept known as the Winters Doctrine. The court held that the tribes had "federally reserved rights" to sufficient water, and that they retained their rights whether they diverted that water or not: the use-it-or-lose-it requirement of the prior-appropriation system didn't apply to them. Subsequent court decisions have established that reservations are entitled to divert enough water to serve all of their "practicably irrigable acreage"; that a reservation's priority date, in a prior-appropriation state, is the date of its creation; and that tribes can sell water rights they don't yet have the infrastructure to exploit. Buzz Thompson, the Stanford Law School professor and Supreme Court special master, told me, "In theory, you have the Indians getting a lot of water, but, of course, theoretical water doesn't help very much, so a large number of the Native American nations have settled their water claims for less than they would probably be entitled to if they went all the way to court."

Determining exactly how much water the tribes are entitled to is known as "quantification"; not surprisingly, Indians have usually been more interested than non-Indians in quantifying Indian water rights. Lined up on the dais behind the podium at the annual meeting of the Colorado River Water Users Association in 2014 were nineteen flags: those of the United States, Mexico, the seven compact states, and ten Indian nations with reservations in the river basin (a group known as the Ten Tribes Partnership). The row of flags made it seem as though the tribes constituted a majority of the participants, but, in fact, they figured in only a tiny part of the agenda, and quite a bit of a speech made by their leader consisted of expressions of gratitude for being included in the program. Still, the full force of existing and pending Indian quantification cases has yet to be felt. It's likely, for example, that essentially all the agricultural irrigation water in Arizona will end up under tribal control, since the tribes' rights are senior even to California's. If that happens, the tribes will in effect become the landlords of virtually all the farmland in the state.

IN 2010, Coconino County suffered a pair of calamities from which it is still recovering: the Schultz Fire, which burned fifteen thousand acres and is believed to have been caused by a careless camper, and the Schultz Flood, which occurred when record rains fell on areas whose covering vegetation had been destroyed by the Schultz Fire. (Both the fire and the flood took place on and around Schultz Peak, in Coconino

National Forest, northeast of Flagstaff.) Flooding seems like an improbable threat in an area where people transport drinking water in pickup trucks, but large parts of the topography of northern Arizona have been shaped, and continue to be shaped, by sudden inundations of vast quantities of rapidly moving water—as you can easily observe. The closer I got to the Colorado, as I drove north from Flagstaff toward Lake Powell, the more the landscape came to be dominated by canyons and washes and creek beds and enormous rounded rock formations— all utterly dry but obviously formed, at some point, by water.

Lake Powell was created by Glen Canyon Dam, which was completed in 1966. It is the largest single element of the Bureau of Reclamation's Colorado River Storage Project, which was authorized by Congress in 1956 and includes dams and reservoirs in all the upper-basin states. The original version of the project included a dam in the Echo Park section of Dinosaur National Monument, on the Green River, a Colorado River tributary, in northwestern Colorado. But the idea of inundating a beloved and spectacularly beautiful conservation area led to protests by a number of groups, including the Sierra Club and the National Park Service. "No one has asked the American people whether they want their sovereign rights, and those of their descendants, in their own publicly reserved beauty spots wiped out," Bernard DeVoto wrote in *The Saturday Evening Post* in 1950. The government eventually dropped Echo Park from the proposal, but the much larger project at Glen Canyon was retained. David Brower, the Sierra Club's president at the time, was an important figure in the Echo Park campaign. He didn't object to the creation of Lake Powell, and he later said that his failure to do so had been "the biggest sin I ever committed."

The main stated purpose of the Colorado River Storage Project is to hold water for use by the upper-basin states, but its biggest beneficiary is really California, because one of the most significant functions of the project's reservoirs is to enable the upper-basin states to meet their lower-basin water-delivery requirements, as established by the Colorado River Compact, at times when their own rainfall and snowmelt are below average. Lake Powell is farther downstream than almost any upper-basin water user except itself, and it could never be tapped, for example, to send water back to farmers in the Grand Valley or to people who live in Denver. Its most important function in recent years has been as a sort of reserve tank for Lake Mead, the biggest lower-basin reservoir. In fact, the Bureau of Reclamation generally manages Powell and Mead as a single system, even though the two are almost three hundred miles apart and are separated by (among other things) the Grand Canyon.

Steadily declining reservoir levels have been the main water concern at Lake Powell for the past decade and a half, but that hasn't always been the case. In 1983, heavy rains and above-average snowmelt in the upper basin forced the Bureau of Reclamation, for the first time, to open the dam's spillways—valve-like control structures that can be used to release excess water—to keep the lake from flowing over the top of the dam. The water shot through the spillways with so much force that large parts of their concrete linings were scoured away, threatening the entire structure. Engineers added temporary barriers to the top of the dam, raising it by several feet, and once the water had receded they repaired the ruined spillways, using a modified design that was intended to reduce cavitation, the force that had ripped them apart. (Cavitation occurs when air pockets form and then implode inside a moving fluid, turning it into a liquid jackhammer.) The repairs were completed shortly before a second huge flood, the following year.

I GOT A LATE START FROM FLAGSTAFF, and as I approached Page, the town closest to Glen Canyon Dam, night had already fallen. For quite a few miles, the sky ahead was dominated by a strange glow and what looked increasingly like enormous searchlights pointing straight up. The source of the glow turned out to be floodlights on the Salt River Project–Navajo Generating Station, a coal-burning electricity plant on tribal land near the southern shore of the lake. The plant is conspicuous during the daytime, too. Its three concrete flue-gas stacks are each 775 feet tall; they are the third, fourth, and fifth tallest man-made structures in Arizona. The towers stand on high ground above the lake, and they and the clouds of combustion gases they emit are visible for many miles in all directions. They are the main source of a filmy haze that's often visible in the sky above eleven nearby national parks and wilderness areas.

The Navajo plant has a complicated history. It was built in the 1970s and was treated at the time as an environmentally less destructive alternative to further damming of the Colorado. The idea that environmentalists could ever have welcomed an agreement to build a facility that burns more than twenty thousand tons of coal a day can seem farfetched to someone wrestling with energy and climate issues in the twenty-first century, but it's a good example of the way perceptions of environmental trade-offs change. Dams are still unloved, because of the harm they do to river systems, but hating them today requires more nuanced thinking than it did thirty or forty years ago. The only noncarbon energy source in the United States that's larger than hydro is nuclear power; solar still barely makes the list.

The land the Navajo Generating Station stands on is leased from the Navajo Nation, but the plant itself is owned by a group of six non-Indian entities, including the federal government and the Los Angeles Department of Water and Power. One of the plant's purposes is to supply electricity to move Colorado River water to Phoenix and Tucson from a diversion more than three hundred miles downstream from the plant and more than three hundred miles west of Tucson. The coal burned by the plant comes from the Kayenta Mine, which is also on

Navajo land and is operated by Peabody Energy, a company that mines coal on every continent except Antarctica. Most of the workers in the plant and the mine and on the seventy-eight-mile-long rail line that connects them are Indians; in 2015 the combined operation generated about a quarter of the revenues of the Navajo Nation and two-thirds of the revenues of the Hopi Tribe. The plant was built where it is mainly to give it easy access to the Colorado River: coal-fired plants don't need as much water as hydroelectric plants do, but they still need a lot, because they generate electricity by turning water into steam and also use water as a coolant.

In 2015, following a five-year negotiation, the Environmental Protection Agency issued a ruling that will allow the plant to continue operating until 2030 as long as it cuts its output by a third by 2020, most likely by shutting down one of its three generators. The ruling will allow the plant to continue operating until 2044 as long as it reduces its nitrogen oxide emissions by eighty percent by 2030. (In 2044, presumably, the plant will shut down for good.) The mayor of Los Angeles, meanwhile, announced that that city's Department of Water and Power would sell its stake in the plant (21.2 percent) by 2019; NV Energy—a Berkshire Hathaway subsidiary, which supplies electricity to much of Nevada—has also promised to sell its share (11.3 percent). The emission reductions required by the EPA should eventually make a noticeable improvement in air quality throughout the region.

THE TOWN OF PAGE has a population of roughly seventy-four hundred. It was founded in 1957 to provide housing for the people who were building Glen Canyon Dam and was originally known as Government Camp. It's even more dependent on the Colorado River than Flagstaff is. Many local businesses are involved in renting, selling, repairing, storing, or provisioning houseboats, which are the principal form of transportation on Lake Powell. The biggest houseboats carry more than a dozen passengers, and the fanciest ones rent for more than $2,000 a day, not including fuel, ice, food, or alcohol. A website operated by Aramark, which is the concessioner of Lake Powell Resorts & Marinas, says, "Imagine driving a golf cart around the world's biggest parking lot—that's how easy it is to operate a houseboat!" You just need to be eighteen years old and have a valid driver's license. "Because houseboats are slow moving," the website also explains, "most guests use their houseboat as 'home base' and choose the thrill of exploring the lake and its coves on a powerboat or personal watercraft. Don't have one? No problem. We have all kinds of water toys to rent—so you can tow them right behind your houseboat."

I drove across the dam and parked in the huge parking lot at Lake Powell Resort, which overlooks the lake. A section of the lot is reserved for RVs, and strewn among its oversize spaces were many rocks, which RV owners had used to chock the wheels of their big machines. The resort used to be closer to the water than it is now; as the lake has shrunk, the paths and boat ramps have had to be extended, some of them a long way. I walked down to Wahweap Marina, directly below the lodge, and snooped around. The two biggest houseboats I noticed had both been built by Bravada Yachts, a company based in Phoenix. It sells a range of models, including some that are more than a hundred feet long and have a helicopter landing pad on the roof. Even the relatively modest houseboats I saw had plenty of amenities: water slides, Jet Ski berths, propane grills, patio furniture, satellite-television dishes, and central air conditioners.

My visit to Lake Powell was in mid-November, beyond the end of the main tourist season, so there wasn't a lot going on. As I wandered among the houseboats, I was followed, in the water, by two ducks and an enormous carp. All three were apparently accustomed to being fed from the dock. I passed a woman walking two pugs, one on a leash and one not.

She was the only resident houseboater I saw. A clerk in the marina store told me that the coming weekend would probably be the last real weekend of the year, and that I was just his third customer that day, not counting marina employees who had used the restroom around the corner. So I bought a Diet Coke and a book, and moved on.

7.

LEES FERRY

If you're in a car and driving west from Glen Canyon Dam, your last chance to cross the Colorado before you reach Nevada is on the Navajo Bridge, which carries Route 89A over Marble Canyon in north central Arizona, twenty miles downstream from the dam. There are actually two bridges side by side: the original, which opened in 1929 and is now used only by pedestrians, and its replacement, which is wider but otherwise similar and was completed in 1995. Both are open-spandrel steel-deck arch bridges—a kind you can make with an Erector set.

I parked on the near side and walked across the pedestrian bridge. Marble Canyon is 750 feet wide at that point, and the drop to the water from the center of the bridge is 500 feet. (A sign says not to jump.) The canyon, like many other topographical features on and near the Colorado, was named by John Wesley Powell, a one-armed Civil War veteran and geology professor, who led the first two successful boat expeditions through the Grand Canyon, between 1869 and 1872. Marble Canyon's walls are actually limestone, but the river had scoured and polished them to such an extent that to Powell they shone like marble.

On the day I visited, the river looked green, sluggish, small, and wholly incapable of making limestone look like anything but limestone. I stood for a long time at the center of the bridge, and took pictures of the canyon in both directions, and gazed at the Vermilion Cliffs—just a couple of miles off, on the far side—and spoke with a man and woman who were doing those same things. In 1928, Glen and Bessie Hyde, a young married couple on their honeymoon, passed under the spot where we were chatting and then under the first Navajo Bridge, which was under construction. Eighty-five miles downstream, they hiked up the South Kaibab Trail and spoke with several people, among them a reporter for *The Denver Post*. Then they returned to the river and were never seen again. Their boat, which Glen had built, was recovered a couple of hundred miles down the river, along with their camping equipment, diaries, and photographs. Their bodies were never found. Had Bessie survived their trip, she would have been the first woman to travel the length of the Grand Canyon in a boat.

No cars crossed the other bridge, in either direction, while I was gazing up and down the canyon, and no boats passed underneath. There's an attractive facility on the western side, the Navajo Bridge Interpretive Center, and I walked over to have a look, but it was closed for repairs, so I returned to my car and drove across the other bridge. Just past the interpretive center and just before the parking lot of the Marble Canyon Lodge, I turned right onto a dirt road, which followed the river upstream. I passed an apartment-building-size sandstone stump called Cathedral Rock and crossed a low bridge over the Paria River, which is a minor Colorado tributary. The road ended in a parking lot near the point that marks the boundary between the river's upper and lower basins as defined by the Colorado River Compact: Lees Ferry. Above Lees Ferry, water mainly flows into the Colorado; below it, water mainly flows out.

LEES FERRY ISN'T A TOWN. It's a scattered assortment of dusty parking lots, mildly interesting informational signs, and tiny, hot-looking government-owned buildings. Its location played a role in western history even before the compact was signed, though, because for many western migrants in the 1800s the canyons of the Colorado constituted a major navigational obstacle, and until the first Navajo Bridge was built the broad river bend near the Paria confluence was the last place for hundreds of miles where crossing the river on a horse or in a wagon was feasible. The fording point was discovered in 1864 by Jacob Hamblin, a prominent Mormon, who had been sent by Brigham Young to scout river crossings. (The Colorado blocked migrant routes from the south and east and stood between the Mormons' Utah settlements and possible expansion beyond the river.) Fording the Colorado there was possible because parts of the canyon walls to the east and south sloped gently enough to permit the construction of a wagon road, and the valley of the Paria, on the other side, provided an exit. The Mormons eventually established a crude ferry service, mainly for the benefit of other Mormons. It operated until June 7, 1928, when Lewis Nez, a twenty-eight-year-old Navajo who worked at a nearby trading post, attempted a crossing with two other men and their Model T Ford. The ferry capsized, and all three men drowned.

Lees Ferry was named for John Doyle Lee, who ran the ferry in the early 1870s. He was a Mormon leader and an adopted son of Brigham Young. (Young adopted a number of his closest associates; Lee sometimes signed his name J. D. L. Young.) Lee was one of the primary participants in the Mountain Meadows Massacre, which took place thirty miles west of what's now the northern end of Zion National Park, roughly 150 miles west of Navajo Bridge. On the morning of September 7, 1857, a detachment of the Mormon territorial militia, called the Nauvoo Legion, aided by a group of Paiute Indians, fired on a group of non-Mormon emigrants who were traveling in a wagon train from Arkansas to California. Seven members of the party were killed, and the survivors drew their wagons into a circle. A five-day siege followed. It ended when Lee, under a white flag, persuaded the emigrants that if they would surrender their arms and abandon their possessions the militia would escort them to safety. Instead, Lee and his men led them a short distance from the wagons, and, on a shouted signal, opened fire and drew their knives, sparing only seventeen of the youngest children. In all, 120 emigrants were killed. Lee distributed the surviving children among numerous Mormon families to keep them from speaking with one another and to reduce their effectiveness as witnesses. Two of the children testified later that, while in Utah, they had seen one or another of Lee's nineteen wives wearing jewelry and dresses that they recognized as having belonged to their mothers.

There was a public outcry in other parts of the country, but federal officials were slow to act. Then the Civil War intervened. In 1871, as the authorities finally closed in, Brigham Young sent Lee to operate a ferry at the Colorado River crossing that Hamblin had discovered. Young was apparently hoping both to protect Lee and to isolate him from people who might ask him questions—and the river crossing was a good place to do that, because it was a long way from everywhere else. (A member of Powell's river crew described the area as "desolate enough to suit a love-sick poet.") Lee was captured by federal marshals four years later, in the town of Panguitch, where he had been visiting one of his wives. He was tried twice and was the only massacre participant ever convicted. On the day of his execution, in 1877, he posed for a photograph while sitting on his coffin and asked that copies be given to his three remaining wives. "Center my heart, boys!" he shouted to the members of his firing squad, who were concealed behind a canvas drape hung from the side of a wagon. Some of Lee's descendants maintain a nonprofit organization and a website devoted to family history and genealogy. He had sixty children, so there's lots of both. Among his descendants are Bradley Udall, who is a great-great-grandson not only of Lee but also of Jacob Hamblin, and Bradley's uncle Stewart Udall, whose middle name was Lee.

The compound where Lee and various members of his family and their successors lived while operating the ferry is Lonely Dell Ranch—supposedly named by Lee's seventeenth wife, Emma Louise Batchelor, who accompanied her husband into exile, gave birth to three of his children there, and ran the ferry on her own for two years after his execution. I found it easily and parked in a small lot at the foot of the path. The site is maintained by the National Park Service, and several buildings that date to a period just after Lee's death are still standing, including a semi-subterranean stone structure that must have been used to keep things cool. I walked around a re-created orchard, which the park service irrigates with water diverted from the Paria and from which visitors in season can pick fruit, and I followed a marked trail to a small, depressing cemetery, where a number of people who lived at the ranch are buried—but not Lee, whose grave is in Panguitch.

JOHN WESLEY POWELL was born in Mount Morris, New York, in 1834. When he was in his twenties, he rowed a boat down the Mississippi River from Minneapolis to the Gulf of Mexico. He served in the Union Army during the Civil War, and lost most of his right arm in the Battle of Shiloh, in 1862. After the war, he was hired as a professor of natural sciences at Illinois Wesleyan University, in Bloomington, despite having never earned a college degree. He led an expedition to Colorado in 1867, and he returned to the West two years later to explore the Green and Colorado rivers. Powell's interest in the region was mainly scientific; his specialties included geology, cartography, and anthropology. The trip also attracted the attention of people who hoped that the Colorado might be exploitable as a migration route to the Far

West—among them Brigham Young. (Cross-country travel was very much in people's minds at that time. The two sections of the transcontinental railroad were joined by the "golden spike," at Promontory Summit, Utah, two weeks before Powell's trip began.) On May 24, 1869, Powell and nine companions set out from Green River, Wyoming, in four wooden boats, one of which Powell had named after his wife. Only two of the boats and six of the explorers made it all the way to the end.

Powell wrote about his expedition in newspaper and magazine articles and, later, in a popular book, first published in 1875 and then reissued, in a new edition, twenty years later. He embroidered parts of his account and mixed in details from a second expedition, taken two years later, but he was nevertheless an evocative observer, and *Canyons of the Colorado*, as the 1895 edition was titled, is still highly readable. "The walls now are more than a mile in height—a vertical distance difficult to appreciate," he writes of some cliffs near the Grand Canyon's beginning. "Stand on the south steps of the Treasury building in Washington and look down Pennsylvania Avenue to the Capitol; measure this distance overhead, and imagine cliffs to extend to that altitude, and you will understand what is meant; or stand at Canal Street in New York and look up Broadway to Grace Church, and you have about the distance; or stand at Lake Street bridge in Chicago and look down to the Central Depot, and you have it again."

To negotiate the more daunting stretches of the river, members of the party had to "line" their boats past them, by attaching ropes to bow and stern and drawing each boat over the rapids from the edge of the river or from a cliff wall, sometimes while two men remained on board to push the hull away from rocks and to keep the rope from snagging on outcroppings. Wallace Stegner, in *Beyond the Hundredth Meridian*, a biography of Powell, originally published in 1954, writes, "Even lining was too dangerous at some of these cataracts. They had to unload, make a trail among the boulders and talus, and carry everything, including the two ponderous oaken boats, stumbling and staggering in hundred-degree heat down to the foot of rapids where as likely as not a careful look showed them another portage directly ahead." And in some places, where the narrowness and steepness of the canyon left them no alternative, they could only point themselves downstream and hope for the best.

Early in the trip, at a point on the Green River that Powell named Disaster Falls, one of the boats was swept over a cataract, dumping her crew and eventually smashing to pieces against the rocks. All of the expedition's barometers, which Powell used to measure cliff heights and other elevations, had been on board, and he was in despair. But the next day the party discovered the boat's cabin, mostly intact, wedged among some boulders downstream, and two of the men were able to recover the barometers. They let out a cheer, and Powell was pleased that the men, who were not scholars, had taken this unexpected interest in the scientific side of the expedition—but then it turned out that what they were cheering about was not the barometers but a three-gallon keg of whiskey, which they had also recovered.

Powell was courageous to the point of recklessness. One day, he became stuck on the face of a canyon wall more than six hundred feet above the river, and because he had only one hand to hang on with he was unable to continue in any direction. One of the men climbed to a ledge just above his head, removed his long johns, and lowered them, like a rope. Powell let go of his handhold, grabbed the long johns, and was pulled to safety. On another day, his companions rescued him from a different canyon wall, four hundred feet above the river, by jamming two of their longest oars into cracks in the precipice, one to keep him from falling backward and one for him to use as a foothold.

Or so Powell wrote in his book. Stegner suspects that the oar incident is fiction, since neither Powell nor any of his men mentioned it in notes they took at the time. But, even if it didn't happen, the trip was spectacularly life-threatening. "Ever before us has been an unknown danger, heavier than immediate peril," Powell wrote. "Every waking hour passed in the Grand Canyon has been one of toil. We have watched with deep solicitude the steady disappearance of our scant supply of rations, and from time to time have seen the river snatch a portion of the little left, while we were a-hungered." The party began with supplies for ten months but lost most of their provisions to accidents and spoilage. They had to strain musty flour through mosquito netting to make it even minimally palatable, and after repeatedly drying and boiling the rancid remains of their bacon they finally abandoned it as inedible. And there were other torments. "They met headwinds so hot and strong they could not run against them," Stegner writes. "Even in mid-river, soaking wet, they panted with the heat, and on shore the sand of their campsite blew over them so that they covered their heads with blankets and sweltered."

Three months into the trip, at the head of an especially fearsome-looking stretch of rapids, three of the men decided they'd had enough. They climbed out of the canyon at a point now known as Separation Creek, intending to walk to civilization. As Powell and the others discovered later, they were killed by Indians shortly afterward, apparently after being mistaken for someone else. Their deaths were especially unfortunate because the rapids they were determined to avoid turned out to be less lethal than they had appeared to be. The rest of the party made it past them without incident and reached the mouth of the Virgin River, and the end of the truly dangerous water, just two days later. Powell, at that point, declared the trip complete. The spot where they stopped is now at the bottom of Lake Mead.

The expedition made Powell famous, and Congress gave him $10,000 to repeat it. For the return, he was determined to avoid the provision problems that had nearly ruined the first expedition. He scouted the area in 1870 and consulted with Brigham Young and Jacob Hamblin. Hamblin turned out to be especially useful as a guide and an intermediary with the Indians. He helped Powell secure a promise of safe passage from the tribe whose members had killed the three men from the first Powell expedition. He also met John Doyle Lee, with whom he enjoyed discussing geology.

Among the precautions that Powell took for the second trip was to cache food and boats at intermediate points. One of the caches was near the Paria crossing that Hamblin had discovered, and as winter approached in 1871 members of his crew left boats and food there. When they returned in the spring, Frederick S. Dellenbaugh—who served on the second expedition as an artist and a cartographer—recognized Lee, who had been sent by Brigham Young not long before. "He was plowing in a field," Dellenbaugh wrote later, in his own highly readable account, *A Canyon Voyage*, published in 1908, "and when he first sighted us he seemed a little startled, doubtless thinking we might be officers to arrest him. One of his wives, Rachel, went into the cabin not far off and peered out at us. She was a fine shot, as I afterwards learned." Lee invited them to dinner, and the meal was prepared by "Mrs. Lee XVIII," whom Dellenbaugh described as "a stout, comely woman of about twenty-five, with two small children, and seemed to be entirely happy in the situation." One member of the party repeatedly made a joke of sneaking up behind Lee and cocking his rifle to watch him jump. A third wife was present as well, but Dellenbaugh never learned her name or number. Dellenbaugh was seventeen when the trip began. Thirty-three years later, he was one of the founders of the Explorers Club in New York City.

Powell was less fully engaged in the second expedition than he had been in the first, and he ended it less than halfway down the Grand Canyon, at Kanab Creek. But his accounts of his discoveries on both trips established his reputation as an explorer and a scientist. In 1879, he was chosen to be the first director of the Bureau of Ethnology of the Smithsonian Institution, a position he held until his death, in 1902. And from 1881 until 1894 he directed the U.S. Geological Survey, where one of his projects was to assess the capacity of the western United States to support irrigated agriculture. In 1893, he was invited to Los Angeles to address the annual meeting of the International Irrigation Congress. When the attendees endorsed an ambitious irrigation scheme that he believed to be unsustainable, he abandoned the speech he had prepared and delivered a warning that no one in attendance was interested in hearing. "When all the rivers are used," he said, "when all the creeks in the ravines, when all the brooks, when all the springs are used, when all the reservoirs along the streams are used, when all the canyon waters are taken up, when all the artesian waters are taken up, when all the wells are sunk or dug that can be dug, there is still not sufficient water to irrigate all this arid region." He was booed, but continued. "What matters it whether I am popular or unpopular? I tell you, gentlemen, you are piling up a heritage of conflict and litigation over water rights, for there is not sufficient water to supply these lands."

AFTER I'D SPENT half an hour or so nosing around Lonely Dell Ranch, I drove back to the main road and then half a mile up the river to the gauging station. On the far bank I saw a box-like concrete pillar, which contains the measuring equipment with which officials of the USGS monitor the river's volume. When they need to check the gauge itself or take water samples from the middle of the river, they climb into a metal cage and ride across on a steel cableway suspended above the water. A measuring station has operated at the site continuously since 1921, and for the past eighty years readings have been taken several times an hour. As a result, the USGS's discharge dataset for the Colorado is almost certainly the most comprehensive such record for any stream anywhere in the world.

On my side of the river was a large parking lot and gently sloping gravel beach, which is the departure point for most modern-day Grand Canyon raft trips: Lees Ferry is still the last convenient entry point to the Colorado River above the Grand Canyon. (The next spot downstream where the river is accessible to a motor vehicle is more than 260 miles away.) A trip was forming when I arrived, and a dozen people, mostly in their thirties, were loading gear into eighteen-foot rafts. They were being watched by an employee of the National Park Service, which controls all rafting trips and licenses their operators. In roughly the same way that scientists re-create historical weather patterns by studying tree rings, you could probably estimate the length of river trips by counting cases of Bud Light. This one was going to last two and a half weeks.

I myself have ridden a raft on the Colorado, but only once and not for two and a half weeks. In 2006, my son had just graduated from high school and my daughter had just graduated from college, and my wife and I celebrated with them by renting an RV in Las Vegas and visiting Hoover Dam, the extraordinary national parks in southern Utah, and the North Rim of the Grand Canyon: the last big family vacation. Our turnaround point was Moab. While we were there, we hiked in Arches and took a one-day raft trip on the "family-friendly" (that is, cataract-free) stretch of the Colorado that flows past the Fisher Towers. The main instruction we received from the raft operator was the same one my mother used to give my siblings and me almost any time we left the house when we were little: go before you go. Bodily wastes don't break down in the desert the way they do in wetter, more organism-rich environments, and the operator told us that, if we had to pee during the trip, we should jump into the river and do it as we swam, and that if we needed to do more than that we would have to be escorted away from the group to sit on something that looked like a surplus World War II ammo can. The weather was perfect, the scenery was remarkable, and the river was gentle. If you got hot, you could roll over the side and swim along for a while. The other occupants of our raft were seventh-graders on a school trip from Orem: blond-haired, blue-eyed girls and boys, some of whom conceivably could have been descendants of John Doyle Lee or of Jacob Hamblin or of members of the Nauvoo Legion.

The rafting group I saw at Lees Ferry had rented their equipment from Ceiba Adventures, an outfitting company based in Flagstaff. Rachel Schmidt, who owns the company with her husband, told me that the river no longer bears much resemblance to the one that Powell traveled down. During both of Powell's expeditions, water flows in the Colorado were highly variable and unpredictable, and there were stretches where the vertical difference between high and low water was a hundred feet. "The river dropped away abruptly," Dellenbaugh wrote of one particularly thrilling stretch, "to rise again in a succession of fearful billows whose crests leaped and danced high in air as if rejoicing at the prospect of annihilating us."

It's not like that now. "The Colorado is a dam-controlled waterway," Schmidt told me, "so we don't have the natural flows that some other rivers have—where you get a big snowpack, and then the snow all melts and you have high water in the spring and early summer. What we have to run on is whatever the Bureau of Reclamation gives us." During recent years, the bureau has occasionally attempted to re-create something like natural flows by briefly releasing a larger volume of water from Lake Powell (named for John Wesley, of course), but there is still always a federal hand on the faucet. Schmidt said that, in some ways, constraining the river had made it more "user friendly" for rafters, by eliminating extremes—the group I saw would have known pretty much exactly what to expect all the way to their destination—but that it also had had major negative impacts on riverine ecosystems. Tom Kleinschnitz, who owns Adventure Bound USA, a rafting company that runs guided trips on the Colorado and two of its tributaries, the Green and the Yampa, told me, "On an undammed river, like the Yampa, the free flows spike up real high. They kind of purge the river corridor and tear out the non-native, noxious weeds, and ten years after a high-flow event you'll see ten-year-old cottonwoods growing in a line where the water reached that year." That doesn't happen anymore on the Colorado, and the fact that it doesn't has contributed to the spread of aggressive invasive plant species throughout the river corridor.

A relatively recent development in the ongoing evolution of the Law of the River is the creation of what are known in Colorado as "recreational in-channel diversions" (RICDs). These are similar to instream-flow water rights, discussed briefly in chapter 3, and can also be thought of to some extent as water rights that belong to the water itself. The prior-appropriation system is based on *diversion*—on removing water from streams in order to put it to approved uses beyond the streambed. In the 1980s, the Colorado Supreme Court, overruling a lower court, held that in certain instances water could legally be considered to have been diverted even if it never left the stream. (The case involved a boat chute and a fish ladder that the city of Fort Collins had built to circumvent a dam.) That decision and similar ones have been controversial, and a number of legal questions remain unresolved, not least because river recreation and species preservation didn't appear on nineteenth-century lists of beneficial uses. There are many potential complications. "Does a kayaking course need 10 cfs or 100 cfs?" Jones and Cech ask in *Colorado Water Law for Non-Lawyers.* "Can the appropriator place a call on upstream junior diverters when there are no kayakers on the course in case someone decides to use the kayak course later that afternoon? If only one kayaker is on the course, is that sufficient to support a call for more water?" Environmentalists have tended to support any ruling that keeps water within riverbanks, but recreation is a broad category, since it includes not just kayaks but also things like houseboats and Jet Skis. And RICDs have the same drawback that instream-flow water rights do: because they're new, their priority dates stink. But their existence, even as theoretical constructs, beneficially enlarges the legal conception of what a river is for.

"The biggest question is whether there's going to be enough water here in the future," Kleinschnitz told me. "We're now at a tipping point. Are recreation and agriculture important, or are we going to do what we did in L.A. and just erase some of those? The decisions we make during the next fifty years will dictate the next thousand." He said he worries that over-allocation and poor water management will eventually make the Colorado wholly unnavigable for recreational users. If that happens, he said, people looking back five hundred years from now will perceive little difference between our era and John Wesley Powell's, and will lump them together as "back when they used to run rafts."

8.

BOULDER CANYON PROJECT

There are two ways to drive from Lees Ferry to Hoover Dam: the northern route, which passes through St. George, Utah, and takes about five hours, and the southern route, which passes through Flagstaff, Arizona, and takes about six. On a map, the two routes together form what looks like a bumpy circle drawn by a nursery schooler. Neither route passes very close to the Colorado, which follows a far more direct path—the one John Wesley Powell took through the Grand Canyon.

I took the northern route—along the foot of the Vermilion Cliffs and past the turnoff for Route 67, which leads to the Grand Canyon's North Rim—and arrived at Hemenway Harbor, near Lake Mead's southwestern corner, late in the afternoon. I parked in a sloping gravel lot, next to a heavy steel cable that ran all the way to the water: it was helping to anchor one of the harbor's docks. Not many years ago, that parking lot would have been underwater: Mead's volume has fallen by just over sixty percent since 1998, the last time it was full, and there are places where its shoreline has receded by more than a mile. From the lot I could see Pyramid Island to the north, and Saddle Island just beyond it; neither is still an island. An earthen causeway connects Pyramid Island to the mainland, and two cantilevered piers extend like wings from its sides. There used to be a "No Fishing" sign at the end of one of the piers, but it hasn't been needed for years. A section of the lake to the south of the causeway was once reserved for scuba divers. Today, you can explore it in hiking boots.

On the dock, I met Bob Gripentog, whose family has owned the Lake Mead Marina since 1957, when he was six, and Rod Taylor, a regional vice president of Forever Resorts, which operates a marina on a different part of the lake. Lake Mead Marina hasn't always been in Hemenway Harbor; it was towed there in 2002 because its original location, in a shallower bay, was rapidly becoming lakefront. Gripentog has lived in Las Vegas all his life. His mother's parents owned a grocery store in town, and for a while his father was stationed at Nellis Air Force Base, at what's now the northeastern corner of the city. Marinas are in his genes, maybe: his sister and her husband also own one, in Kentucky.

We got into one of Gripentog's boats and went out to explore. The dying light made the surrounding terrain look like rumpled mountains of exotic spices: all ochers and umbers and oranges and smoky taupes and rusty browns. Both Gripentog and Taylor said they were concerned about what Taylor described as "apocalyptic reporting" about water in the West, and wanted to be sure I understood that the lake is still gigantic. He characterized Mead's surface area with a unit of measurement that people in the Northeast seldom have opportunities to use: three Disney Worlds, about eighty-seven thousand acres. He said that he hoped potential visitors would not be deterred by news reports about shrinkage. "As you can see," he said, "we have a lot of water out here."

That's all true, and if you're weighing a vacation you shouldn't let media hysteria keep you away. Lake Mead National Recreation Area was visited by more than six million people last year, making it one of the most popular destinations in the National Park system. Still, if you know Mead at all, you can't help noticing that most of it is missing. In fact, the lake today contains only about thirty-eight percent as much water as it did in 1998. The loss is easy to visualize because as the lake recedes it exposes a white "bathtub ring" on the surrounding bluffs, created partly by minerals in the water and partly by leachates from the rock. Gripentog steered his boat around a promontory and into a cove-like finger of the lake which wasn't visible from the marina. As he did, we passed close to a canyon wall, the lower portion of which was as white as chalk. The lake and the surrounding landscape are so vast that when you see the bathtub ring from far away you have little sense of the scale. Viewed up close, though, it makes you gulp: the distance from the surface of the water to the top of the white band that day was 130 feet. Then we cleared the promontory, and suddenly, straight ahead, was Hoover Dam. A line of buoys warned us to stay back, well away from the intakes that draw water from the lake into the turbines in the power plant on the other side.

I first saw Hoover Dam in 2003, while taking a break from the World of Concrete, a trade show I was covering in Las Vegas. I arrived at the visitors' center in time to take one of the day's last tours. From the guide I learned that the dam is more than 700 feet tall, more than 1,200 feet wide at the top, and roughly 660 feet thick at the base, and that it's made almost entirely of concrete—approximately 3.4 million cubic yards of it, or roughly 7 million tons, with an additional million cubic yards in appurtenant structures. After my tour had ended, I walked onto the top, toward Arizona, and leaned over the belly-high parapet on the downstream side. From there, the face of the dam plunges down and away, and it must engender dark yearnings in the minds of certain kinds of skateboarders. Despite everything I had heard and seen and read about the dam until that moment, it was only as I stood on its rim and gazed down toward the river far below that I gained a full, vertiginous sense of the extraordinary pile of concrete at my feet—the "callous, cruel lump," in the words of a visitor in the 1930s. I felt the same sense of mild unease I felt once as I floated on my back in the deepest part of a deep lake and imagined the unsettling volume beneath me.

Three years later, I visited Hoover Dam again, with my family on the first day of our post-graduation western trip. We drove our rented RV right over the dam and parked in a lot on the Arizona side. (The two-lane road on the dam was closed to civilian vehicle traffic in 2010; you now cross the canyon on a similarly breathtaking bypass bridge, a little way downstream.) I felt the same sense of excitement I had during my first visit, and I urged my wife and children not only to gaze over the side but also to use the public restrooms built into the parapet—perhaps the most beautiful dam-based public restrooms in the world, with their polished brass doors and Art Deco fittings and beautifully designed and executed terrazzo floors. When we reached the visitors' center, I pointed out the two thirty-foot-tall winged bronze statues next to the flagpole, and explained that their black diorite pedestals had been assembled by lowering their components onto blocks of ice, so that, as the ice melted, the installers could easily make micro-adjustments in their position without scratching their highly polished surfaces. Inside the visitors' center, my son took a photograph of me standing slack-jawed in front of a diorama depicting the operation of the cableway that was used to place the concrete during the construction of the dam, and another of me reading an information sign that asked "Why Was Hoover Dam Built?" (Answer: "The Colorado River is both friend and foe. It has the power to sustain life and ruin lives, to create opportunity and destroy prosperity.") I learned later that, while I was absorbed in these wonders, my wife was whispering to our daughter, "Don't worry—it won't all be like this."

. . .

CONSTRUCTION OF HOOVER DAM was anticipated in the negotiations that led to the Colorado River Compact. California, in particular, needed a big lower-basin reservoir to decrease the river's downstream silt load and to reduce the danger of catastrophic flooding in the Imperial Valley. In 1928, Congress passed the Boulder Canyon Project Act—a major piece of water legislation, and the most expensive appropriation bill in U.S. history to that point—which authorized the construction of a dam that would be twice as tall as any previously built. The project was too ambitious to be managed by a single contractor. The winning bid was submitted by a consortium called Six Companies, which later also built Parker Dam and the Colorado River Aqueduct (and which actually included seven companies, two of which entered the project as partners). Before the dam was authorized, engineers from the Bureau of Reclamation had determined that a site which seemed promising originally, in Boulder Canyon, was unsuitable; Black Canyon, a few miles downstream, was substituted, although the name of the project wasn't changed. The dam was known both as Boulder Dam and as Hoover Dam virtually from the beginning; Congress made the current name official in 1947.

The first major phase of the project was at least as daunting as the building of the dam itself: the boring of four enormous diversion tunnels through the canyon walls, two on each side of the Colorado, so that the entire river could be piped around the construction site until the dam was finished. The rough borings were fifty-six feet in diameter and averaged about four thousand feet long. To create them, as many as thirty workers at a time stood or crouched on truck-mounted carriages called "jumbos," each of which had four scaffold-like tiers, and operated rail-mounted hydraulic drills that looked like anti-aircraft guns. In all the old photographs I've seen, nobody is wearing ear or eye protection, even though rock chips flew everywhere and the noise was so loud and so continuous that workers had to communicate with hand signals. Ten dedicated blacksmith shops used oil-fired furnaces and enormous sharpening machines manufactured by Ingersoll Rand to recondition the drill bits, of which there were thousands. (No ear or eye protection in the blacksmith shops, either.) Once a section had been completed, the jumbo backed up, miners packed the drilled holes with dynamite and gunpowder primers, and the charges were set off. Then muckers removed the shot rock, with assistance from a power shovel and a line of waiting trucks, and the routine began again. Crews worked around the clock in eight-hour shifts, almost always racing against each other, and by the time they had perfected their technique they were able to advance at an average rate of a foot or two an hour. The completed rough borings were lined, all around, with a three-foot layer of concrete—100,000 cubic yards per tunnel—yielding a finished interior diameter of fifty feet.

Jobs in the tunnels employed as many as fifteen hundred men at a time. They were extraordinarily unpleasant and dangerous, and not only because of the noise. Even by the standards of paleo-capitalism, Six Companies' commitment to human decency was low. There were just two days off a year, Christmas and the Fourth of July; workers could be fired on a whim; the temperature in the tunnel was sometimes as high as 140 degrees; and contractors padded their profits by cheating employees through innumerable instances of what Edmund Wilson—who visited the site in 1931, on assignment for *The New Republic*—described as "systematic skimping, petty swindling, and frank indifference," including overcharging for food and lodging. If anything resembling modern labor laws had been in effect, Six Companies could have been prosecuted as a criminal enterprise.

Inside the tunnels, the air was thick with smoke and nitroglycerin fumes, and there were cave-ins, rock slides, misfires, vehicle accidents, electrical short circuits, and major equipment malfunctions, many of which caused injury or death. Joseph E. Stevens, in *Hoover Dam: An American Adventure*, published in 1988, writes, "To visitors touring the construction site, one of the most memorable—and terrifying—sights was the procession of empty muck-hauling trucks racing backward down the winding, precipitous grades into Black Canyon, the drivers standing up in the open cabs with one foot planted firmly on the accelerator and the other on the running board, craning over their shoulders to see where they were going"—a strategy that saved a couple of minutes per trip by eliminating the need to turn around. One of the greatest perils inside the tunnels was carbon monoxide poisoning. The tunnels were inadequately ventilated, and gasoline powered trucks and bulldozers operated inside them and often sat for long periods with their engines idling. This was known to be a lethal practice, and in tunnel construction elsewhere electric-powered vehicles had become virtually standard. But electric vehicles were expensive, and Six Companies was determined to hold down costs by using a single truck fleet project-wide. Michael Hiltzik, in *Colossus: Hoover Dam and the Making of the American Century*, published in 2010, writes that, for workers, a sign of rising carbon monoxide concentrations was that "the electric lights in the tunnel acquired a bluish halo." When a worker succumbed, men known as "chasers" were supposed to quickly drag his unconscious body into the open air for possible revival. Because carbon monoxide poisoning left no physical trace, doctors employed by the project, without exception, assigned causes of death that relieved Six Companies of the obligation to pay compensation—usually pneumonia or cardiac arrest. In 1931, Six Companies won an injunction preventing Nevada from enforcing a state law that prohibited the use of gasoline engines in mines, and a federal court eventually ruled that the law didn't even apply, because a diversion tunnel isn't a mine. Besides, by then the tunnels were nearly finished.

Yet whenever a worker died, was mangled, or became too sick to work, thousands of aspirants were ready to take his place. Some jobs paid less than fifty cents an hour, and no one was paid for the time it took to be transported to the site—which could be considerable—but jobseekers were so plentiful that Six Companies actually cut some wages in 1931. Workers struck, but the strike was broken easily. Jobs were scarce everywhere, and unemployed men and their families, including pregnant women and infants, arrived daily from all over the country—a difficult trip, since Black Canyon was essentially wilderness. (It was where James Christmas Ives, who led an Army expedition up the river from Mexico in the late 1850s, had to abandon his principal transport, a sternwheeler riverboat, which had broken up on submerged rocks.) Virtually all the workers were white. Only a tiny number of black men were hired, and only after protests by the NAACP and others. (The Great Depression was especially hard on minorities, since jobless whites were willing to do work that in better times they had considered beneath them.) Black workers were employed only in the gravel pits, which were even hotter than the canyon, and they were required to drink from separate water buckets, and they were conveyed to and from the site on a segregated bus. There were no Asians in the workforce, because the federal construction contract forbade the use of "Mongolian labor." That ban originated with the Chinese Exclusion Act of 1882, which prohibited the employment of workers from China, and it remained in the charter of the Bureau of Reclamation until 1944. Even so, the summertime heat in Black Canyon was so intense that a writer in 1931 observed that some people had "prophesied that Orientals would finally have to be imported to cope with the melting temperatures."

Between 1930 and 1931, the population of Las Vegas tripled, from five thousand to fifteen thousand. The dam contract required Six Companies to build a camp large enough to house eighty percent of its workforce, using federal money and federal land, but completing the camp, called Boulder City, took almost two years. Until it was ready, some unmarried workers were housed in a sweltering barracks in the canyon itself. Others slept in Las Vegas, on newspapers in front of the courthouse or the Union Pacific depot. Many of those who had arrived with wives and children lived in improvised shelters in a squatters' camp known as Ragtown, in Hemenway Wash, near the spot where Bob Gripentog's marina is anchored today. During the summers, the temperature in Ragtown often reached 120 degrees during the day and didn't fall below 100 degrees at night. Night-shift workers had trouble sleeping during the heat of the day, and often managed no more than what Edmund Wilson described as "a heavy sweating coma." The air was so hot and dry that workers who owned cars had to park them in the river before driving to town for supplies or medical care, so that the cars' wooden wheels and spokes (standard at the time) would swell sufficiently to remain intact for the round trip. Sanitation was prehistoric. Water from the river was undrinkable until it had sat long enough for most of the silt to settle out, and the unprotected tanks in which it was stored became reservoirs of dysentery. The young wife of one worker became so ill during the hottest part of the summer of 1931 that she attached a plea for help to the collar of her German shepherd and pushed the dog out of her tent, but by the time aid arrived she was dead. She was one of four women who died in Ragtown that day.

Jobs outside the tunnels were also brutally difficult and dangerous. The canyon's walls were prepared by "high scalers," who used jackhammers to remove loose rocks and outcroppings while dangling hundreds of feet above the riverbed in primitive wooden bosun's chairs, which looked like children's swings. The sun made the canyon walls so hot that touching them with bare skin could be agonizing—Wilson described summer winds in the canyon as "a furnace-breath"—and rubble knocked loose by men working high in the canyon often fell on men working below. Still, high scalers were widely admired and were celebrated for their bravado. One day, an engineer from the Bureau of Reclamation slipped during an inspection, and a high scaler working twenty-five feet below heard him cry out and saw him begin to slide down the precipice. "Without hesitation," Stevens writes in *Hoover Dam*, "he pushed out from the wall, propelled himself horizontally through the air, swung back in to the cliff face, and snagged the falling man's leg." He held the engineer in place, upside down, until two other high scalers could reel him in.

Because no one had ever built a dam as big as Hoover Dam, no one at the outset was certain how big it needed to be. Like the Empire State Building—another first-of-its-kind supersize structure, the construction of which also began in 1930—the dam was overbuilt by a large margin, and its size made some people fear that its mass and that of the reservoir behind it would throw the earth off its orbit. Concrete for the dam was manufactured in a large batching plant that was built for that purpose on a ledge near the bottom of the canyon, about two-thirds of a mile upstream from the site. When each new load was ready, a worker threw a lever, dumping it into an eight-cubic-yard steel bucket on the back of a railroad flatcar parked below. A locomotive moved it along a precarious-looking trestle that protruded from the canyon wall like a shelf, and at the dam the buckets were picked up one at a time with a 650-foot-long cable—which was attached to an aerial cableway suspended across the top of the canyon—and swung to where they were needed. The dam was cast in interlocking rectangular columns measuring between twenty and forty feet on a side. New concrete was added to the tops of the columns in layers five feet deep, and in a staggered pattern so that each new section could harden before more concrete was added on top or to any side.

The engineers devised their construction techniques as they went along. Workers known as "puddlers" stood inside the forms as each load was dumped, then spread the new concrete with long-handled shovels and (mainly) their feet. Buried within each new section were two separate networks of pipes: one that would later be used to inject portland-cement grout into cracks, joints, and gaps between columns, and one that would be used to cool the concrete as it cured, by circulating refrigerated water through it. Concrete doesn't "dry"; when you add water, the portland cement that holds it together undergoes an exothermic chemical reaction, "hydration," which causes it to form crystals that powerfully lock together and gain in strength indefinitely. As the cement hardens, it heats up and expands, and as it cools again it contracts. The engineers had realized that a Hoover Dam–size mass of concrete, during the early stages of hydration, would generate and trap so much heat that without artificial cooling the structure would require more than a century to cool and would develop spectacular cracks as it did. As each section of the dam reached the desired internal temperature, measured by thermometers also buried in the concrete, the cooling pipes serving it were disconnected, drained, and filled with grout. A persistent rumor about Hoover Dam is that the bodies of workers killed during construction are also embedded within it; the guide who led my tour in 2003 assured us that that isn't true, but he said that there are plenty of bodies buried in other places.

THE BOULDER CANYON PROJECT didn't lure only the unemployed. Even before it was finished, the dam had become a major tourist attraction, and its appeal as a travel destination was enhanced by the legalization of gambling in Nevada in 1931. The first casinos were built mainly to draw dollars away from dam workers, but tourists liked gaming, too, and by the mid-1930s the dam and the blackjack tables were annually attracting hundreds of thousands of visitors, initiating the otherwise unforeseeable transformation of Las Vegas from a remote desert outpost into a major city. "Union Pacific was running excursion trains to Boulder City almost weekly," Stevens writes, "and on March 9, 1934, the railroad's special, ultramodern streamlined train, carrying celebrities, members of the press, and company and government officials, actually glided into the canyon and up to the face of the dam on the contractors' construction line."

The project, in a somewhat less direct way, also fueled the growth of Los Angeles. Visitors to Hoover Dam often assume that the electricity produced by its power plant goes entirely to Las Vegas, and especially to the Strip, but in fact Nevada receives only about a quarter of the output. Roughly sixty percent goes to metropolitan L.A., 270 miles to the west, and one of its functions there is to power the pumps that move water from places that have it to places that don't. Beginning in the late 1930s, electricity from Hoover Dam added impetus to the suburbanization of the Southern California desert. And, since greater Los Angeles was the principal source of early visitors to Las Vegas, people and money surged back in the other direction, too—a potent feedback loop.

Generating electricity was a minor consideration for California when it first began pushing for a lower-basin dam and the reservoir, but it ended up being the key to federal approval of the Boulder Canyon Project, because it was by selling electricity that Congress expected everything to be paid for, over a period of fifty years. Hoover Dam has a "nameplate" generating capacity of more than 2,000 megawatts, at which rate it produces roughly 4.5 billion kilowatt-hours a year. As Lake Mead has shrunk, however, that capacity has shrunk, too, by about a quarter so far, because the output of a hydroelectric plant is largely a function of the quantity of water passing through its penstocks—which at Hoover Dam are thirty feet in diameter—and the elevation difference between the intakes and the turbines. (The power plant's biggest year ever was 1984, an extremely wet, high-water year, during which it

produced more than 10 billion kilowatt-hours.) The relationship between lake level and output isn't one-to-one, though, because once the
force of the falling water has dropped below some critical threshold,
cavitation and other destructive effects inside the turbines can cause
their electricity production to plunge or become erratic. It seems likely
that if the situation became truly dire Lake Powell would be sacri-
ficed to keep water flowing through Hoover Dam, but if that happened
hydroelectric generation at Glen Canyon Dam would be lost. (The
minimum "power pool" level for Lake Powell is roughly a hundred feet
below where its surface is right now.)

No one knows the exact elevation at which the Hoover or Glen
Canyon power plants would truly fail, because the effects aren't perfectly predictable and the lakes haven't been as low as they are today
since the years when they were filling. Taking them or any other large
hydroelectric facility entirely offline would represent a significant set-
back for non-carbon energy production in the United States—like
winding back the clock on solar and wind—both because of the direct
loss and because most of the replacement plants would inevitably be
powered by fossil fuels. Water issues are never only about water.

LAS VEGAS

After we'd gotten as close as we could to Hoover Dam, Bob Gripentog turned his boat around and we headed north, along the lake's eastern shore. The sun was very low. Gripentog's glasses had photochromic lenses, and by now they were almost untinted. To the west we saw an Air Force helicopter, black against the darkening sky, hovering above the lake. Its rotors were fanning the water. They created a low cloud of fog-like mist, which extended surprisingly far in all directions. Every so often, a tiny figure would descend on a rope, dip into the water, then rise back up: a chilly training exercise. We looped to the left, back toward the marina, and passed Saddle Island. Near its southern end was a steel tower a little taller than the bathtub ring, and standing on top of the tower was a structure that, from a distance, looked like a house with a gable roof. "That was the original intake that sent water to Las Vegas," Gripentog said.

And there, in the minds of many people, is Lake Mead's problem. The decline of the lake and the growth of the city have been concurrent events, so there's a widespread assumption that the latter is the cause of the former, and the association is strengthened by the fact that, of all the cities that draw water from the Colorado River, Las Vegas is the closest to its banks. Particular indignation is directed at the man-made lake in front of the Bellagio hotel and casino—a water feature that contains twelve hundred computer-controlled "dancing fountains" and more than forty-five hundred lights, and is said to have cost $40 million to create. The Bellagio fountains are often cited as a prime example of the sort of wanton waste that got the West into water trouble in the first place.

But all this is wrong. When the Colorado River Compact was negotiated, Nevada had a population of only about eighty thousand, and Las Vegas barely existed. The exact division of water among the three lower-basin states wasn't settled until 1963, but even then Las Vegas was relatively tiny, with a population of less than ninety thousand. Nevada's compact share was set at 300,000 acre-feet, roughly two percent of the paper-water total, and at the time that seemed like plenty, even to people in Las Vegas. Nevada had begun drawing modest amounts of water from the river, by way of the lake, in 1941, at the behest of the federal government, which had established a manufacturing company near Boulder City to produce magnesium for the war effort. Those withdrawals increased in the 1970s, after the completion of a bigger intake, but even today Nevada takes less river water than it's entitled to. The Southern Nevada Water Authority, which controls water use in the region, has implemented some of the most stringent conservation measures in the United States, and today ninety-three percent of the water that's used indoors in the Las Vegas area is treated in a plant to the east of the city and is then either used again locally, mainly for landscaping, or returned to the lake, through the Las Vegas Wash, earning the state a "return-flow credit." The SNWA also has a long-running "xeriscaping" program, which pays people to replace turf with desert plantings and has been copied in other western cities. Residents are fined if they water yards and gardens on days when they're not supposed to or if they allow water to run onto sidewalks or into the street, and commercial users, including golf courses, have to comply with strict water budgets or pay escalating penalties. Las Vegas today uses less water overall than it did fifteen years ago, even though the population of the metropolitan area has grown to nearly two million. A spokesman for the authority told me that, if Nevada were to withdraw a full year's worth of its Colorado River entitlement in a single gulp, the lake would fall by forty-five inches. "For context," he continued, "California uses about fifty-four feet of water a year. Arizona uses thirty-five feet. Mexico uses eighteen and a half feet. To put it in even more context, the evaporation off Lake Mead alone is two and a half times our annual usage as a community."

As for the Bellagio, the pool containing the fountains covers about eight acres and is roughly eight feet deep, giving it a volume of sixty-four acre-feet. That water doesn't come from the lake. It comes from a deep aquifer, which the hotel has a long-standing permit to tap, plus storm-water runoff from the hotel property. (The well permit belonged originally to the old Dunes hotel and its golf course, which no longer exist.) The fountains look big, but as water uses go they don't actually amount to much. If their entire contents were pumped into Mead instead of fired into the air, the surface of the lake would rise by less than a hundredth of an inch. "When people see pictures of the fountains, they angrily jump up and down," an SNWA hydrologist told me in 2014, "but those fountains are extremely popular with tourists, and they definitely bring an economic benefit to the area." How you feel about that will depend on how you feel about the economic vitality of Las Vegas—a city whose modern existence depends on the availability of cheap jet fuel and the inability of the average person to understand basic arithmetic—but worrying that the Bellagio is killing Lake Mead, or even imperiling the natural aquifers underlying Las Vegas, is like believing that unplugging your cell phone will reverse global warming.

. . .

LAS VEGAS MEANS "THE MEADOWS"; it was named for an oasis a few miles north of what's now the Strip. Indigenous peoples lived near the oasis for millennia, and travelers on the Spanish Trail stopped there to water their thirsty herds, and Mormon settlers and others began grazing livestock there on a permanent basis in the mid-1800s. Groundwater in the meadows was at first so abundant that it pushed up through the surface of the desert, forming earthen domes called "spring mounds." That water was much higher in quality than the water in the river—which Edmund Wilson described in 1931 as "an opaque yellow like coffee with too much cream"—and during the first two years of the Hoover Dam project tank trucks transported fifty thousand gallons a day from springs and wells in Las Vegas to Black Canyon, a round trip of sixty miles. At the construction site, workers distributed the water in canvas bags. The deliveries ended only after Six Companies had completed a large facility to clarify, filter, soften, and chlorinate river water sufficiently to make it potable, although even then it was lower in quality than the groundwater in town.

The springs that created the spring mounds dried up in the late fifties or early sixties. They had been fed by shallow groundwater, which originated as precipitation in mountains to the west and north. That water flowed downhill and under the valley, and it broke to the surface when its forward movement was obstructed by buried deposits of fine-grained sediments. It was physically isolated from deeper groundwater by an impermeable stratum beneath it, like the floor of a swimming pool, and for that reason hydrologists refer to it as "perched." Because the perched water occupied a zone that extended no more than a couple of hundred feet below the valley floor, it was drawn down fairly quickly once people began digging wells. That's why the springs stopped flowing.

In 2007, the Las Vegas Springs Preserve, a 180-acre "non-gaming cultural and historical attraction," opened roughly where the oasis had been, on land owned by the Las Vegas Valley Water District—one of seven municipal agencies that make up the SNWA. The preserve contains a botanical garden, hiking trails, performance spaces, archaeological sites, and the Origen Museum, which is devoted to local history. (The name is derived from "original" and "generations," according to the website.) I visited the day after my boat ride with Bob Gripentog. Inside the museum's rotunda I saw a re-created spring mound, which roughly duplicates a real one you can see alongside one of the hiking trails outside, and a simulated desert flash flood, featuring real water. Elsewhere in the museum, I learned that the Colorado River at first "seemed to defy efforts to develop the desert land," that early-twentieth-century residents of Las Vegas used the wooden cover of a water-settling tank as a dance floor during Sunday picnics. I also learned that annual water usage by metropolitan Las Vegas peaked, at 176 billion gallons, in 2007. The decline since then, by 15 or 20 billion gallons, is partly the result of the city's conservation measures and partly the result of the economic implosion of 2008, which hit Las Vegas especially hard. (Recessions are generally good for the environment; so far, they've been the most broadly effective tool, worldwide, for reducing carbon output.)

The preserve's botanical garden covers 110 acres and contains more than twelve hundred species of "native and desert-adapted plants"; it's the largest collection anywhere of cacti and succulents indigenous to the Mojave Desert. Many of the specimens, including dozens of mature trees, were collected from nearby areas where the desert was being bulldozed to make way for residential subdivisions, and in the early 2000s Las Vegas had an abundance of those. I visited the city on a reporting assignment in 2009, and my guide during much of that trip was a young woman who worked in the public information office of the SNWA. She told me that she had first seen Las Vegas in the 1970s, during a vacation with her parents. The area's explosive population growth hadn't begun yet, and her family's first reaction was to wonder where all the hotel and casino employees went when their shifts ended. "We thought maybe they lived in the hotels," she said, "because we didn't see any houses." No one visiting Las Vegas today would have the same misapprehension. If you approach the city from the northeast in a car at night, the lights of its sprawling suburbs make the entire valley look like the glowing caldera of an enormous volcano.

People who look closely at the specimen plants in the botanical garden at the Springs Preserve are usually surprised to see that even the cacti are irrigated: there are little black plastic water emitters poking out from the sand and gravel at their bases. This is true not just of the botanical garden but of most of the city's xeriscapes. The SNWA's turf-removal program, Water Smart, requires that converted areas be planted with replacement vegetation in such a way that the canopies of the plantings, at maturity, will provide ground coverage of at least fifty percent. The purpose of the rule is to ameliorate Las Vegas's airborne-dust problem, which is sporadically severe, and to reduce the ground's absorption and re-radiation of solar energy—the so-called urban heat-island effect, which can raise air temperatures and even alter local weather patterns. "The old mantra for desert landscaping in Southern Nevada, when I first started, was a cactus, a rock, and a dead-cow skull," Patricia Mulroy—who at the time was the general manager of both the SNWA and the Las Vegas Valley Water District—told me in 2009. "What we want is desert landscaping that has a single root system, but can spread all over the ground. That keeps the temperature of the rocks down—which is necessary because in the summer it gets pretty hot around here."

But there are downsides, and the main one is that the non-turf plantings usually have to be watered, too—a requirement that affects most xeriscaping efforts. Dale Devitt, a professor in the school of life

sciences at the University of Nevada at Las Vegas, told me, "The popular idea is that if you remove turfgrass you're going to save unbelievable
amounts of water, but the reality is that there are trade-offs. Removing
the turfgrass is one thing, but if you don't control what goes back in,
and just plant trees instead, within a period of time there's no savings
at all. We've demonstrated, for example, that one mature oak tree requires as much water as sixteen hundred square feet of low-fertility
bermuda grass. People will sometimes remove turfgrass but leave fifty-
foot-tall trees behind, without realizing that the trees were totally dependent on the irrigation that the turfgrass was receiving."

This very issue has arisen recently in California, where homeowners
who have responded to the drought by disconnecting their sprinkler
systems and replacing their flower gardens with desert plants have
sometimes been distressed to discover that their trees, which they had
always thought of less as irrigated vegetation than as permanent landscape features or pieces of outdoor furniture, had gone the way of their
old lawn. I've visited Los Angeles many times, and my son lives there
now, and I'm ashamed to admit that until recently it had never occurred to me that the citrus trees I saw growing in people's yards were
anything but the lush natural flora of Southern California. I now
understand—and it should have been obvious to me long before—that
Los Angeles, like Las Vegas, was built in a semi-arid environment, and
that the vast majority of the plants that grow there, including not just
the citrus trees but also every one of the city's familiar palm trees, were
imported from somewhere else and wouldn't exist without irrigation.

Yet trees are more than just water consumers. Jim Lochhead, the
CEO of Denver Water, told me, "A great deal of the landscaping we
have is not only a social amenity but also an economic driver. If we got
really serious about water conservation and lost our tree canopy, that
would be literally hundreds of millions of dollars in lost property values
across our region. And if we lost our trees we'd lose our protection from

heat-island effect, and that would mean millions of dollars a year in increased energy costs. We get pushed, as water utilities, to simply use less water, but it's not as easy as that." There's a grassy quadrangle near the center of the UNLV campus that's a good example of the same thing. It's shaded by tall trees, and as a result it feels many degrees cooler, and attracts many more lingering students, than the paved expanses beyond its perimeter.

The SNWA's Water Smart program is intended mainly for homeowners, who are among the world's most clueless irrigators, fertilizers, and pesticide appliers, but some of the most enthusiastic participants have been golf courses. Angel Park Golf Club, a thirty-six-hole public facility six miles west of the Springs Preserve, removed seventy-six acres of grass, replaced much of the turf on its driving range with pinkish, pea-size gravel, turned off a fountain, and eliminated three lakes. (Water features, because of evaporation, can require more irrigation than fairways do.) In 2009, the superintendent of another golf course told me that his annual water bill was $1.5 million, even after removing turf. "We do about thirty-six thousand rounds of golf a year," he said. "If you calculate that out, it means that every golfer who heads down our first fairway represents forty-two dollars in water costs." Because that fixed expense places so much pressure on margins, the club's grounds crew constantly monitors soil moisture, using electronic sensors, and can adjust irrigation levels, sprinkler head by sprinkler head. Mulroy told me, "People love to beat up on golf, but what very few people realize is that golf courses have the most sophisticated, high-tech irrigation systems possible, and, as a result, they are the most efficient irrigators in the valley."

There are obviously limits. In 2004, the city imposed a moratorium on new golf course construction. That decision not only prevented the mindless proliferation of irrigation-dependent turfgrass but also improved the ability of existing courses to withstand the subsequent

recession and to absorb the rising cost of water and other critical resources. Whether you think that's a good thing or a bad thing depends
partly, again, on what you think about the economic vitality of Las
Vegas. But even a desert area can support some number of thirsty outdoor recreational facilities, including parks, swimming pools, and ath-
letic fields of all kinds, as long as the costs and trade-offs are identified
and managed intelligently. Golf courses are easy targets—who needs
'em?—but the truth is that most of the cherished activities of most
Americans are environmentally problematic, to say the least.

AS THE POPULATION OF LAS VEGAS GREW, the perched zone that
had once fed the city's springs partially refilled in some areas with water
that originated not in precipitation in the mountains but in human
activities, especially the irrigation of turf. Today in Las Vegas there's so
much groundwater of that type—which Andrew Burns, the division
manager for water resources at the SNWA, told me "really has nowhere
to go"—that almost all large construction projects are affected by it.
There are big buildings downtown, including the Mandalay Bay hotel
and casino, whose owners have to run underground pumps constantly
to dewater their foundations.

I had heard that the water in the Bellagio fountains comes from the
same system, but Burns said that the perched water is too salty to be
used in that way. The sediments that the perched water lies in are naturally high in soluble minerals, especially toward the eastern side of the
city, and the water that seeps into those sediments is salty to begin
with, since most of it comes from the Colorado. As a result, the perched
water is "really sort of nuisance water," Burns said, and if the Bellagio
used it in its fountains visitors would be put off by various unpleasant
characteristics. "It has an aroma to it," he said. "You wouldn't want to
spray it up into the air."

I asked him whether the perched water couldn't be used for something. "We studied that back in 1996," he said. "The idea at that time
was to extract it and blend it with potable water, but economically it
didn't make sense. The water would have to be treated, even to use it on
turf, and the sediments it lies in are so tightly packed that you can't
extract it easily. More recently, we did a pumping test at the Springs
Preserve, where we were trying to find a water supply for some fish
tanks, and what we found was that, even at very low pumping rates,
it was hard to get that water out of the ground with a reasonable yield
and without excessive drawdown. We monitor the perched system, because we don't want it to contaminate our potable groundwater supply,
which is more than a thousand feet deeper, but using that water, so far,
hasn't been practicable." (Since the mid-seventies, the SNWA has maintained the volume of the lower aquifer by withdrawing potable water at
approximately the rate at which it is replenished by precipitation in the
mountains—about sixty thousand acre-feet a year.)

Some of the perched water actually does get used. The buildings on
the Strip that dewater their foundations have discharge permits that
allow them to, in effect, flush the extracted water into the region's
wastewater treatment system, where it's processed along with water
from toilets and kitchen sinks. Some of that treated wastewater is returned to the lake, by way of the Las Vegas Wash, and some is used
directly for landscape irrigation, including golf courses—whose con-
version to recycled water has been an element of Southern Nevada's
water-conservation efforts. Angel Park Golf Club—the course I visited
near the Springs Preserve—was irrigated with potable water for the
first decade after it was built, and during an especially dry year in the
mid-1990s it used 650 million gallons. Not long afterward, the Las
Vegas Valley Water District built a wastewater recycling plant a short
distance from the course, and the club connected to the new main and
built a reservoir across Rampart Avenue from the clubhouse. When I

visited the course, the superintendent took me over to see the reservoir. He splashed his hands near the intake valve to show me there was nothing scary about recycled water, which, he said, was clean enough for bathing. The club's annual SNWA water budget today is a little less than 320 million gallons, or about half its peak consumption, and its efficiency efforts, including turf removal and sprinkler-system modifications, have been so effective that he usually doesn't have to struggle to meet it. Recycled water is becoming increasingly important in other parts of the country, too. Sand Creek Station, an eighteen-hole public golf course in Newton, Kansas, irrigates with effluent water from a wastewater treatment plant owned by the city, which obtained a $450,000 federal grant to build a supply line to the club. That course performs an environmentally valuable public service, by giving the city's wastewater a final "polishing": Sand Creek's fairways and greens help to filter out remaining impurities as the water is returned to the ground.

Irrigating with recycled water is tricky, though, because wastewater treatment doesn't remove salt—which is dissolved and therefore doesn't settle out the way silt does. Dale Devitt, the biology professor I spoke to at UNLV, told me, "Colorado River water naturally carries a ton of salt per acre-foot, and reuse water carries almost twice that, somewhere between one and a half and two tons per acre-foot. A typical golf course uses six and a half acre-feet of irrigation water per acre per year. So, if there's no leaching, they're applying ten or eleven tons of salt per acre per year. That means that after five years, if there's still no leaching, they've added fifty tons of salt to every acre of soil." The same thing happens in yards, parks, and farms that irrigate with recycled wastewater. The accumulating salts harm plants directly and also create a number of more subtle problems by gradually clogging air and water channels in the soil, reducing the ability of roots to absorb nutrients and promoting previously unfamiliar plant diseases.

This is not a new problem. Over a period of roughly eight hundred years, beginning around 600, the Hohokam Indians built and maintained what was then one of the world's largest and most advanced irrigation systems, in what's now southern Arizona. The Hohokam diverted water from two Colorado River tributaries, the Gila and the Salt, and fed it into extensive networks of canals and ditches, with which they irrigated tens of thousands of acres of agricultural land roughly where metropolitan Phoenix is today. The main canals—which the Hohokam built over many generations, using the crudest of tools—were enormous and were engineered with a degree of hydraulic sophistication that, to some people, suggests the involvement of alien beings.

Then, around 1450, the Hohokam and their complex agricultural society disappeared. Exactly what happened is still a mystery to anthropologists, but the leading theories include the cumulative effects of irrigation itself. The Gila and the Salt, like the Colorado, both contain high levels of dissolved salts, and when saline water is applied to crops the salt can build up in the soil and in the groundwater that underlies it, eventually becoming lethal to plants. Shepard Krech III, in *The Ecological Indian: Myth and History*, published in 1999, writes, "Perhaps the best-known example is ancient Mesopotamia, where siltation of canals and salinization of lands watered by the Tigris and Euphrates have been linked to the decline of civilization. . . . In desert regions like Mesopotamia and the Sonoran Desert, farmers cannot count on rain to leach out accumulated salts. Irrigation water must do. Yet the more one irrigates, the more the water table rises, and with surface evaporation, even more salts are left in topsoil." If the concentrations become high enough, agriculture becomes impossible unless something is done to purge salt from the root zone.

"Salinity is one of those things you won't notice at first, because the changes are very subtle," Dale Devitt told me. "But at some point you cross a threshold, and it's downhill from there." To prevent such disasters, irrigators who use recycled water, including farmers, almost always have to irrigate more heavily—a practice that undermines efforts to conserve. Even plant species that are promoted as salt-tolerant have a finite tolerance for saline soil. Robert N. Carrow, a professor of crop and soil sciences at the University of Georgia, told me, "You know the salt flats in Utah? They're devoid of plants for a reason."

PATRICIA MULROY has had a huge impact on water management in the West. She was the principal architect of Nevada's most innovative efforts to acquire and conserve water—including the creation of the SNWA, through which a number of separate municipalities worked together to finance, build, and operate water infrastructure they couldn't have afforded individually. One of her guiding principles has been "shared sacrifice," a notion that, in many ways, runs counter both to the Law of the River and to the narrowly parochial outlook of traditional water managers. She retired in early 2014, after twenty-five years, but remains one of the country's most visible and influential water experts. She is now a member of the staff of Brookings Mountain West at UNLV, where her title is senior fellow for climate adaptation and environmental policy.

I went to see Mulroy at UNLV shortly after my visit to the Springs Preserve. I parked in a lot on the east side of the campus, next to an area that used to be lawn. Workers had removed all the turf and were in the process of replacing it with pinkish gravel, native plants, and young trees, plus a few boulders. There were also a number of existing trees. One section hadn't been covered with gravel yet, and workers there were laying subsurface irrigation lines: hose-like black tubing that would distribute water to an emitter next to each new plant and to soakers arranged in broad concentric circles around the base of each tree. (Coyotes sometimes cause problems with emitters by treating them as chew toys.)

I spoke to Mulroy in her new office, which she appeared to be still moving into—very possibly because she is in so much demand as a speaker and conference participant that she doesn't spend a lot of time there. She has short gray-blond hair and icy blue eyes, and when she talks about water she sometimes looks at you with an expression that I can only describe as fierce.

"Lake Mead is scary right now," she said. "I see it suffering the consequences we've all known for at least ten years it was going to suffer. It was never a question of if; it was a question of when. The amount of water we thought we had in that river system doesn't exist. You can't fault anyone in the 1920s for their assumptions; that was the best science they had. But today we know better, and Mead is going to be stressed for a long time. Does that mean we can't have a series of severe weather patterns that would start bringing it back? Sure we can. But I think this is the new normal."

One of the paradoxes of resource management is that successful conservation programs often depend on population growth—a force that pulls in the opposite direction. When I visited the SNWA the first time, in 2009, Mulroy was dealing with what she described to me then as "a catastrophic decline" in the connection charges that homeowners and businesses had to pay to be hooked up to the water system. The recession and the mortgage crisis had reduced consumer demand for water—a beneficial outcome, you would think—but they had also halted new construction throughout the valley, and without new construction the SNWA's revenues from connection fees had fallen by ninety percent. Rapid population growth had been both the cause of and the solution to Las Vegas's water challenges.

Ordinary efforts by consumers to cut back on water use can have the same counterintuitive effect. "My old finance guy *hated* conservation," Mulroy told me in 2014, "because it bled revenues. He called it the Evil Empire of Conservation. I-pay-you-not-to-buy-my-product is not what they teach in Econ 101. And there are communities all over the United States that are dealing with the same issue, because their conservation efforts have been too successful." Californians made that discovery in 2015, when their water bills rose even as they cut back their use. A spokesperson for the Association of California Water Agencies told a reporter from Reuters, "Droughts are costly for water agencies. Revenues are being affected by the mandatory conservation but at the same time costs are going up." Mostly for that reason, conservation incentives in Las Vegas have always been aimed at consumptive uses—and for homeowners that means exclusively water that's used outside. ("Grass is the ogre.") In other words, no rebates for putting bricks in toilet tanks. Mulroy continued, "Because we recycle so much of our indoor wastewater, we weren't going to gain any water resources by helping people change out the plumbing in their house. People would argue that when they used less water indoors it cost us less to pump water to them. Oh, goody. But that's not the issue."

The decline in revenues that Mulroy worried about in 2009 threatened not only popular initiatives like the Water Smart program but also the construction of Southern Nevada's "third straw"—a new intake in Lake Mead that the association was building to provide its members long-term access to Nevada's Colorado River allotment. The third straw, including a three-mile-long tunnel between the intake and the shore, was going to cost more than $800 million to build, Mulroy told me, and more than half of that amount was originally budgeted to be raised from connection charges. In 2009, she faced the challenge of covering the cost almost entirely from existing users, since growth had come to a halt, and to do that the SNWA had to raise rates. But it did so successfully—one of Mulroy's most impressive skills has always been an ability to build public support for costly ventures—and the third straw was completed in 2014. At roughly the same time, the association received approval to begin building a $650 million pumping plant, also financed entirely by rate payers, to make use of the new intake.

JUST AS PROXIMITY MAKES people think that Las Vegas is the principal cause of the decline of Lake Mead, it also makes them think that any further decline in the lake will be a problem mainly, or even only, for Las Vegas. But that isn't true, either. When the pumping plant for the third straw is completed, Nevada will be the only lower-basin user with the infrastructure required to draw lake water from below the level known as "dead pool"—roughly nine hundred feet above sea level, the elevation of the lowest openings in the four intake towers on the upstream side of Hoover Dam. Approximately a quarter of the water remaining in Mead is below that dead-pool line and, therefore, untappable by users below the dam. The chance that the lake will drop that far anytime soon is small—it's more than 180 feet below the current surface—but fifteen years ago few people thought the lake would ever drop to where it is today.

"If the lake ever gets to those levels," Mulroy said, "it will mean that the upper basin has suffered dramatically, through years of untold drought. And they'll get hit first, because they don't have the reservoir capacity to carry them through anything like that." But the biggest impact will be in the lower basin, because if Lake Mead ever reaches dead pool the amount of water that Arizona and California will be able to draw from the river will drop to zero. "If Mead falls to nine hundred," Mulroy continued, "nothing goes downstream from Hoover Dam. All the reservoirs below it will be bone-dry—Mohave, Havasu, Imperial. But Southern Nevada will still be taking water out of the lake, because the new intake is at eight-sixty"—860 feet above sea level, 40 feet below the lowest Hoover Dam intake. "That's the reality," she continued. "I don't care what your water right is. If the lake goes that low, your water physically can't get to you. You know? Frame that water right. Hang it up on your wall. Admire it. It's useless."

Even though Mulroy, more than anyone else, is responsible for the construction of the third straw, which effectively guarantees Nevada's access to the river in perpetuity, she is not a water isolationist. For years, she has been the most vocal and creative advocate of cooperation among all the states that depend on the river, since she believes that a disaster for one is a disaster for all. "There are still those who want to talk about water winners and losers," she told me, "but they don't understand the interconnected economy of the river. We have plumbed the Colorado to bleed water in all directions. We take water in Wyoming—outside the river's watershed—and move it to Cheyenne. Come down to Colorado: we move it across the Continental Divide, from the West Slope to the Front Range, into the Kansas-Nebraska basin—outside the watershed of the Colorado. We move it across the Utah desert to the Wasatch Front, to Salt Lake, Provo, Orem, and all those agricultural districts—not in the Colorado watershed. In New Mexico, we move it to Albuquerque, which straddles the Rio Grande. In Arizona, we move it across 360 miles of desert, to Phoenix and Tucson and still more agricultural districts. And in California, we move it over hundreds of miles of aqueduct, from Lake Havasu to the coastal cities—not in the Colorado watershed."

All those far-flung places, Mulroy said, nevertheless constitute a single system, which extends far beyond the river itself and adds up to more than a quarter of the economy of the United States. "We may be citizens of a community, and a state, and a country, but we are also citizens of a basin," she said. "What happens in Denver matters in L.A. What happens in Phoenix matters in Salt Lake. It's a web, and if you cut one strand the whole thing begins to unravel. If you think there can be a *winner* in something like that, you are nuts. Either we all win, or we all lose. And we certainly don't have time to go to court."

10.

COLORADO RIVER AQUEDUCT

About two hours south of Lake Mead, on the Arizona side of the river, I arrived at London Bridge. It stood on the River Thames from 1831 until 1968, when London, which was planning to replace it, sold it to Robert McCulloch, an American entrepreneur and chainsaw manufacturer, for $2.5 million. McCulloch hired a crew to take it apart and number the stones, and then he shipped it to Arizona by way of the Panama Canal and Long Beach, California. He reassembled it on Lake Havasu, a Colorado River reservoir on the border of Arizona and California, as the anchor attraction of a development he was creating, Lake Havasu City, which today has a population of fifty-three thousand. The bridge has five graceful granite arches, and it spans a man-made two-hundred-yard-wide channel that separates the eastern shore of the lake from what used to be a peninsula. Whether McCulloch got a bargain or was ripped off is hard to say; I drove over the bridge twice, on my way to and from the resort where I spent the night, and, although I never felt that Pip Pirrip might suddenly pop up from behind one of the railings, the whole thing did seem interestingly out of place—as did the replica of England's Gold State Coach (built in 1762, during the reign of King George III) in the lobby of the London Bridge Resort, at the eastern end of the bridge.

Lake Havasu marks the beginning of the final and most complex stage of the transformation of the Colorado River from a natural stream into a dispersed and brachiating resource-distribution system. At the lake's southern end, some water is diverted west, to Southern California, and some is diverted east, to central Arizona, and some continues downstream, to diversions farther south. The lake was created in 1938 by the construction of Parker Dam, a graceful concrete curve roughly 750 feet from end to end, topped by a blocky colonnade. The website of the Bureau of Reclamation describes Parker Dam as "one part of a system of storage and diversion structures built by Reclamation to control and regulate the once unruly Colorado River," but it was really built to provide water for metropolitan Los Angeles, nearly 250 miles to the west.

Parker Dam and Lake Havasu are both mainly the work of William Mulholland, who created most of L.A.'s early water-related infrastructure. Mulholland was born in Ireland in 1855. He ran away from home in his early teens, eventually joined the British Merchant Navy, had numerous outsize adventures, and ended up in California in 1877. He held a number of water-related jobs and worked for a while on a crew that was laying an iron water pipeline in Los Angeles, which was still barely large enough to be considered a town. He became an American citizen in 1886. A year later, he became the superintendent of the Los Angeles City Water Company, which was privately owned, and five years after that he was hired to run the water company's publicly owned successor, the Los Angeles Department of Water and Power. At his direction, Los Angeles quietly bought land and water rights in the Owens Valley, more than two hundred miles to the north. The LADWP then built an aqueduct to divert water from the Owens River to the city. Farmers in the Owens Valley sabotaged some of the infrastructure, in a revolt known as the California Water Wars, but Mulholland and the city prevailed. The 1974 movie *Chinatown* tells a highly fictionalized version of essentially the same story. In the movie, elements of Mulholland's personality and professional career are divided between two characters: the murdered water-department engineer Hollis Mulwray (good Mulholland) and the creepy sociopath Noah Cross (bad Mulholland, played by John Huston). *Chinatown* takes many liberties with historical fact, but it makes the city's transformation from desert to urban oasis both emotionally and economically comprehensible. (It's also one of the greatest movies of all time.)

Mulholland was a firm believer that the highest and best use of any body of water was to provide economic benefits to human beings. In the mid-1920s—in a conversation recounted by Marc Reisner in *Cadillac Desert*—he told an employee of the National Park Service that the government ought to hire a dozen of the world's best photographers and pay them to spend a year living in the Yosemite Valley and making pictures, in all seasons, of its natural wonders. And when the year was over, Mulholland continued, "I'd go in there and build a dam from one side of that valley to the other and *stop the goddamned waste!*" This conversation, as recounted in the book, seems too on-the-mark to be entirely trustworthy, but it does demonstrate the widely held early-twentieth-century conception of "conservation," which Mulholland shared.

In the early 1920s, Mulholland realized that the water Los Angeles was drawing from the Owens Valley was insufficient to keep pace with the accelerating growth of the city and of the irrigated farms that surrounded it, and he decided that the most promising supplemental source was the Colorado River. To get water from the river, though, he needed both a dam to impound it and an aqueduct to carry it west. Parker Dam, like all dams on the Colorado, belongs to the United States and is managed by the Bureau of Reclamation, but the MWD paid nearly the full cost; it also built the Colorado River Aqueduct, which carries the water west. The entire construction project took eight years, employed thirty thousand people, and coincided almost exactly with the Great Depression. Mulholland died in 1935, three and a half years before everything was finished.

PARKER DAM IS BIGGER THAN IT LOOKS, because it extends 230 feet below the original riverbed, making it the deepest dam in the world. (The engineers had to dig that far to hit solid rock. What they were digging through was silt deposited by the river.) When I reached it, thirty miles south of London Bridge on Route 95, I answered a few security-related questions from uniformed guards, then drove across the top and followed a winding road up into the hills.

The Colorado River Aqueduct begins at an enormous pumping plant on the lake's western shore, roughly two miles above the dam. The plant was named for William Paul Whitsett, who was the chairman of the MWD from 1929 till 1947. At a locked gate outside it, I met Donald Nash, who manages the desert portion of the aqueduct. He was wearing sunglasses and a black-and-orange Harley-Davidson T-shirt, and he had a photo identification tag clipped to his belt. He gave me a tour, during which we were joined by his daughter, Baily, a high school junior, who was staying with him for the weekend. Running the dam and the other facilities used to require more people than it does today, and when Baily was younger there were so many children in residence that they filled the local elementary school, whose sports teams were the Parker Dam Rams. At that time, a number of families connected with the dam lived in a small mobile-home village a short distance downstream; the mobile homes are gone now, but you can still see the crumbling streets and streetlight poles. The Nashes lived in a smaller cluster of mobile homes much closer to the power plant. Being an aqueduct manager is less lonely than being a lighthouse keeper, but not by a huge margin. Nash said, "I don't know if you've noticed, but there isn't a lot to do around here." Still, Baily said she had loved growing up next to the power plant. When I arrived, Nash guessed correctly that she would be inside, drawn by the smell of oil in the pump room.

The plant's main building has a terrazzo floor and Art Deco light fixtures, and it contains nine 9,000-horsepower General Electric pumps with robin's-egg-blue housings. "That's eighteen school buses per pump," Nash said, as we walked down the line. "Each one could fill an Olympic-size swimming pool in twenty seconds." The pumps run so smoothly that when you place a nickel on any of them—as visitors are sometimes encouraged to do—there isn't enough vibration to make it slide off. They take Colorado River water from the lake and push it through nine enormous pipes, which rise three hundred feet up a steep slope directly behind the plant. From the top of the slope the water flows through a mile-long tunnel to Gene Wash Reservoir, in the Whipple Mountains. A second pumping station then pushes it higher still, and through a six-mile-long tunnel, to Copper Basin, a bigger reservoir. Nash said, "Then it goes by gravity down to Iron Mountain and Iron lifts it 144 feet; then to Eagle, and Eagle lifts it 438; then to Julian Hinds, and Hinds lifts it 441." Nash turned to his daughter. "Am I getting it right so far?" he asked. She said he was. Altogether, there are five pumping stations, 92 miles of tunnels, and 147 miles of open aqueducts, buried conduits, and siphons. The siphons are minor masterpieces of early-twentieth-century hydraulic ingenuity: they carry the water, without mechanical assistance, under desert washes, to protect the aqueduct from inflows of silt during flash floods. "They're like the P-traps in your house," Nash said. "The water comes in on one side and daylights slightly lower on the other."

The open parts of the aqueduct are fenced and guarded. The tunnels are roughly sixteen feet in diameter, and every February cleaning crews scrub them with machines that look like tractors covered with bottle brushes. "Five guys ride on each one," Nash said. "They go down in the hole and put out the brushes, and move forward at two miles an hour, and if the sides of the tunnel look bad they make multiple passes." The same crews drag the open parts of the aqueduct year-round. Cleaning can yield eight inches of additional clearance—enough to make a significant increase in the aqueduct's carrying capacity. In an average year, the system moves 1.2 million acre-feet, or four times the Colorado River allocation of the state of Nevada.

By the terms of its original construction agreement with the federal government, the MWD receives half the electricity generated by Parker Dam. That's not nearly enough to move all that water all the way to Los Angeles, so the district buys additional electricity on the spot market. The usual practice at the plant is to run the pumps all out until both reservoirs are full, then shut down until the reservoirs need to be filled again—a practice that saves roughly $30,000 a month in electricity costs, Nash said. To cut its electricity costs further, the plant has a load-sharing agreement with Los Angeles, and shuts down its pumps in four-hour blocks if, for example, L.A. needs extra current to run residential air conditioners.

Lake Havasu's water volume doesn't fluctuate the way Mead's and Powell's do; the intakes are close to the surface, and the level of the reservoir is maintained at all times within a band of just a few feet. Water takes about five days to travel from Havasu to Lake Mathews, forty-five miles southeast of downtown Los Angeles, and there it enters a distribution system that covers much of the metropolitan area. Recently, because of the drought, some of the water has been pushed even farther, to suburbs well beyond downtown L.A. "I actually went over and watched them start that pump," Nash said. "It was kind of exciting to see my water going all the way to Ventura."

. . .

WE WALKED OUTSIDE. The pump building protrudes a short way into the lake, and a terrace with a decorative concrete railing runs all the way along the front, on the water side. "If you throw a line into the water here," Nash said, "you're pretty much guaranteed to catch a bass." Farther out, people fishing from boats sometimes snag submerged trees, remnants of the days before the valley was flooded to create the lake. People used to swim and run their boats right up to the plant—the pumps create no perceptible current, and a swimmer can't be sucked in—but after 9/11 a security zone extending about a hundred feet into the lake was established around the plant, marked by a rope and red plastic floats. Nash was instructed to install a bulletproof barrier in front of the plant's enormous transformers, which date to the 1930s and are a short distance from the building, but at his suggestion he was allowed instead to hide them behind a cluster of tightly spaced trees—a more aesthetically pleasing screen.

Nash lifted a metal grate in the terrace to show me a problem that he and his crew have been dealing with for the past half-dozen years. Under the grate was a steel coupling of some kind, but I can't describe it more precisely because it was thickly covered by what looked like wet gray rock-wool insulation. "Those are quagga mussels," Nash said. "In the middle of the summer, they'll have built up so bad that you won't be able to squeeze your hand through the grating at the bottom of the trash rack." Quaggas are an invasive species and are closely related to zebra mussels. They are indigenous to parts of Ukraine, and they probably got to the United States in the bilge water of oceangoing ships. They were first noticed in the Great Lakes in 1989 and in Lake Mead in 2007—a territorial leap whose magnitude surprised scientists—and have now been found along the entire length of the Colorado, from

Grand Lake down. Nash picked up a few shells from the ground and arranged them on the palm of his hand. All were roughly dime-size. "This is an Asian clam," he said, pointing to a shell that was almost round. "It's invasive, too, but we're used to it. These over here are quaggas—the ones that look more like fingernails. They multiply, multiply, multiply, and they love stagnant water." They cluster in dead spaces in submerged parts of the plant's innards, including drains, trash racks, and heat-exchanger pipes. Divers come twice a year to blast the dam's intake gratings with a high-powered underwater pressure washer, and every other year a crew dredges up huge piles of shells and hauls them away in a dump truck that Nash and his crew have modified for that purpose.

"Quaggas don't like heat, and they don't like chlorine, and they don't like copper, and they don't like silicone," he continued. "I have a test box down here on the end that we've coated with a paint that's got copper in it—kind of like anti-fouling paint on a sailboat. We suck in water there with an extra cooling pump, and they don't like it. And there's a really bright blue silicone coating that they also hate, but you can scratch it off with a fingernail." Mainly, they use chlorine. "We can't do it here," Nash continued, "because this is a federal waterway, and we can't do it in the reservoirs, because they're full of fish, so we chlorinate at the outlet of Copper Basin, where the aqueduct begins, using a twelve-and-a-half-percent solution. It's about ninety percent effective." Quaggas have no natural predators in this part of the world, and they can be carried from one body of water to another by fish, birds, bait buckets, anchors, boats, and boat trailers. Metropolitan L.A.'s newest reservoir is Diamond Valley Lake, about fifty miles west of Palm Desert, completed in 2003. It's quagga-free so far, and MWD is trying to keep it that way by no longer routing Colorado River water into it. (The lake is fed by the Inland Feeder, an in-state system whose ultimate sources are in Northern California.)

We got into a big pickup truck, and Nash drove us back down to the main gate, on Intake Pump Camp Road. The road is narrow, and for more than a mile it winds along the side of a steep slope directly above the lake. (When Baily's mother drove it for the first time, after Nash got the job, she said, "Tell me there's another way to get to our house." There isn't.) We passed an employee picnic area in a little oasis beside a stream, which is fed by water oozing through sandstone formations on either side of the dam at Gene Wash Reservoir. Light rain was falling, and members of the local Girl Scout troop were setting up tents in a grove of palm trees near the edge of a large swimming hole. The stream empties into the lake at Gene Wash Cove, and Nash said that on Labor Day weekend the cove becomes so crowded with boats and inebriated celebrants that you could almost walk from one side to the other without getting your feet wet—one of the few times each year when people who work at the facility don't feel that the lake belongs just to them.

We passed Gene Village. It contains a second pumping station, which moves water from Gene Wash to Copper Basin, as well as mobile homes, a guest lodge, a mess hall, a swimming pool, tennis courts, and a museum. We passed a mile-long airstrip, which Nash, who has a pilot's license, sometimes uses when he needs to visit the aqueduct's other pumping stations, each of which has an airstrip of its own. "Back before I became the boss, this was the funnest road to drive on," he said. "Now I have to drive the speed limit." We passed the turnoff for Black Meadow Landing, a low-cost lakefront resort, mainly for people with RVs. We passed through an automated security gate. The area surrounding Copper Basin is enclosed by a tall chain-link fence to keep trespassers away from the reservoir. Nash said that two hikers had climbed over the fence recently but hadn't been able to get back out and had used the phone at the security gate to call for help. We passed some huge piles of quagga shells—the remnants of dredging operations. We passed a tree containing a bald-eagle nest, which Nash said hadn't housed eaglets for the past two years—most likely because golden eagles, which are bigger, had been raiding it. We passed a small house on the shore of the reservoir, with a boat on a trailer parked in front.

"Russell Ingram lives there," Nash said. "He's the patroller, grader operator, dam inspector—whatever you want him to be—and he runs people off if he sees them. There's a weir below the dam, and he takes readings there to make sure the runoff hasn't increased and the dam is still sound. That keeps our insurance rates low. He's got a one-of-a-kind septic system, and he's got his own domestic water system, which is extra unique because it's big enough to serve as fire suppression for the house." Ingram is married, and he and his wife have another house, far outside the compound, but he spends most of his time living by himself at Copper Basin, which for him is like a private fishing, boating, hiking, and motorcycling preserve.

We stopped at the far end of the reservoir, where water at that moment was flowing into the aqueduct at the rate of 1,720 cubic feet per second. The tunnel opening is less than twenty feet below the surface, meaning that much of the reservoir's volume is dead storage. Near the outlet was the reservoir's quagga control system: a pair of silo-like storage tanks filled with sodium hypochlorite—basically, household bleach. An injector adds six gallons to the aqueduct each minute, and the tanks are refilled by big trucks, which make three deliveries a day.

On our way back out, we ran into Ingram. He came over to Nash's truck and leaned on the driver's-side mirror. "I saw a good-size bighorn sheep yesterday," he said, "but it's been so wet that they haven't had to come down from the mountains." He was topping off the fuel tanks of his fleet of big trucks, including a grader. A major storm was expected that night—in the distance we could already see dark clouds that looked almost like special effects—and he needed to be ready to clear the road if rain washed it out, to make sure the chloride tankers would be able to get through. Maintaining the road is made difficult by state environmental restrictions: he can't simply use his backhoe to scoop sand from the desert, because doing that might damage the habitat of a protected tortoise species. When he needs a new supply, state officials will fence off a small area to keep tortoises out, and he can then take material from within the enclosure and stockpile it at designated spots along the roadway. Nash invited him to join him and Baily that evening for pizza, beer, and a campfire in front of Nash's mobile home, weather permitting, and Ingram said that he would be there. "It's not a good Harley day," he said.

11.

CENTRAL ARIZONA PROJECT

On the southeastern shore of Lake Havasu, on the other side of Parker Dam, is the beginning of Arizona's equivalent to the Colorado River Aqueduct: the Central Arizona Project, a 340-mile-long system of canals, tunnels, and pumping plants that extends all the way to Tucson. It wasn't completed until 1993, and the main reason is that it took Arizona and California a long time to work out critical water-related disagreements.

Arizona didn't become a state until 1912, just ten years before the negotiation of the Colorado River Compact. (It was the forty-eighth state—the last before Alaska and Hawaii, half a century later.) Its population was still tiny in 1922, and its water consumption was small, but its negotiators believed that they had at least as much right to draw water from the Colorado River as California did, and that their state's potential for growth would be severely limited if it didn't receive a full share. This was one of the main points of contention that Herbert Hoover addressed when he suggested that the river states agree to divide the Colorado into two basins, leaving the individual allocations to be settled later. But Hoover's compromise sidestepped the problem rather than solving it, and by 1928 none of the participating states had ratified the compact. That same year, Congress passed the Boulder Canyon Project Act—the bill that authorized the construction of Hoover Dam, as well as some other infrastructure, which California, especially, was interested in. But the act came with a condition: California and at least five of the six other participating states had to ratify the Colorado River Compact within six months of the bill's passage. (Arizona was known to be a likely holdout.) In addition, California had to agree to an annual allotment of 4.4 million acre-feet.

The Boulder Canyon Project Act also authorized the lower-basin states to permanently apportion their half of the river by specific amounts: 2.8 million acre-feet for Arizona and 300,000 acre-feet for Nevada, in addition to the 4.4 million acre-feet for California. All those numbers were close to ones that the upper basin states had proposed to the lower-basin states during negotiations a year before, and eventually they did become the official allotments—but Arizona again objected. Its issue was not the numbers themselves but what they were meant to include. The compact treats the Colorado River and all its tributaries as a single system; that's why Wyoming and New Mexico, which don't actually touch the mainstem but do contain rivers that contribute to it, were parties to the negotiations. The Colorado does flow through Arizona, but Arizona also contains a tributary, the Gila River, which joins the Colorado not far from Arizona's border with Mexico. The Boulder Canyon Project Act granted Arizona "the exclusive beneficial consumptive use of the Gila River and its tributaries within the boundaries of said State," but didn't specify how much of that water, if any, would be counted against Arizona's proposed 2.8 million acre-foot share under the compact. California thought the Gila's entire flow should be deducted, leaving Arizona with as little as 500,000 acre-feet from the river itself; Arizona thought none of the Gila's flow should be deducted, or, if not none, then much less than all. As a consequence, Arizona's legislature refused to ratify the compact. The six other states did ratify it, though, and California agreed to the 4.4-million-acre-foot share, and Hoover Dam and the other projects went forward.

Congress hadn't required the upper-basin states to set their own allocations, because at that time they drew very little water from the river, and how they divided it among themselves was of little significance to anyone. When they did set their shares, in 1948, they used percentages rather than acre-feet—a method that Congress and the others states would have been wise to employ in the lower basin, too. (The division the upper-basin states agreed to was 51.75 percent for Colorado, 23 percent for Utah, 14 percent for Wyoming, and 11.25 percent for New Mexico.) Among other benefits, allocating water that doesn't exist is much less likely to cause trouble later if you don't specify exact amounts.

WHILE HOOVER DAM was under construction, California began building the Colorado River Aqueduct and Parker Dam. Arizona's governor, Benjamin B. Moeur, viewed the dam as an act of theft. Like many Arizonans, he worried that Southern California would suck the river dry before Arizona was in a position to divert almost any of its own share, whatever that turned out to be, so he sent a small National Guard detachment to the construction site to make sure that neither the workers nor the dam touched land on the Arizona side of the river—a challenge for a dam builder, you would think. The National Guardsmen borrowed a small ferryboat from Nellie Trent Bush, a state legislator who lived in the town of Parker, a few miles downstream. As the boat approached the site, it became entangled in a cable attached to a construction barge, and the National Guardsmen had to be rescued by their putative enemies, the people working on the dam. Moeur later sent a message to President Franklin Delano Roosevelt in which he said that he had "found it necessary to issue a proclamation establishing martial law on the Arizona side of the river at that point and directing the National Guard to use such means as may be necessary to prevent an invasion of the sovereignty and territory of the State of Arizona." By that time, his National Guard detachment had grown to include many more soldiers, as well as a number of trucks with machine guns mounted on them. Moeur also made Nellie Bush "Admiral of the Arizona Navy."

Nellie Bush was born in 1888 in northern Missouri, in a tiny town not far from the tiny town where my father's father was born, at roughly the same time. When she was five, her father moved the family to a tent village in Mesa, Arizona, in the hope that the dry desert air would relieve his tuberculosis. (In the era before antibiotics, Arizona marketed itself as a haven for pulmonary patients of all kinds. Scottsdale was sometimes called White City, because so many residents lived in white-canvas tents.) She married in 1912. Three years later, she and her husband bought a two-boat ferry business in Parker, and she became the first woman to earn a pilot's license on the Colorado River. The Bushes' usual cargo was ore—copper, gold, and manganese, all of which were mined nearby—along with the automobiles of a growing number of self-propelled tourists.

Because no upstream dams had been built yet, navigation on the river could be exhilarating. "Waves sometimes would be eight feet high," Bush recalled later. "Often when we were caught on the river in a storm, we'd have to throw overboard some of the ores. Many a time when the sailing was dangerous and I thought about my baby in the pilot house, I've uttered a little prayer, 'Now if you'll just let me get this kid off of here alive, I'll never bring him back on board again.' But you forgot about that after the danger had passed." A banker cheated her once, so she decided to become a lawyer, and in 1924, when she was thirty-six, she received a law degree from the University of Arizona. She said years later that she and the one other female law student in her class had been asked to leave the room during a discussion of a rape case. "I asked if they had ever heard of a rape case which didn't involve a woman," Bush said. "They let us in after that." She practiced law for many years and served multiple terms in the state legislature. She earned an airplane pilot's license in the 1930s, when her son became interested in aviation. "I realized that as a mother I could retain my son's interest, only as long as I could speak his language," she said. She died in 1963.

When Moeur declared martial law, Harold Ickes, the secretary of the interior, halted work on the dam and referred the matter to the Supreme Court—which surprised Ickes and nearly everyone else by ruling in Arizona's favor, on the grounds that the Boulder Canyon Project Act had not specifically authorized Parker Dam. Congress passed an authorization bill a few months later, and construction resumed. The dam was completed in late 1938, and water first flowed to Los Angeles in 1941. The aqueduct's success made California seem like an even bigger threat to Arizona than it had before, in part because creating a similar diversion on the other side of the river was going to require an even more ambitious project than the one that California had just completed, and there was no possibility that Arizona would be able to finance it on its own. Before seeking federal help, though, Arizona had to accede to the terms of the Boulder Canyon Project Act by ratifying the Colorado River Compact. And in 1944, mostly for that reason, it did.

Arizona's fundamental disagreements with California remained unresolved, however, and eight years later it sued. The Supreme Court—which tries lawsuits between states—appointed Simon H. Rifkind, a prominent corporate lawyer in New York City, as special master. (Rifkind was actually the case's second special master; the first died in

1954.) He conducted hearings in San Francisco for two years, heard testimony from more than two hundred witnesses, and collected four thousand exhibits. Arizona began by arguing that the court ought to ignore the prior-appropriation doctrine and divide the water equitably, but it later changed its strategy (and its lawyer) and argued instead that its priority right to make beneficial use of all the water in the Gila River had been perfected well before 1922 and was therefore beyond the reach of the compact. That argument prevailed. In December 1960, Rifkind issued a 433-page report to the Supreme Court in which he sided with Arizona on all issues. In 1962, the court itself heard sixteen hours of oral arguments, and the following year it published a 5–3 decision in which it accepted Rifkind's recommendations, giving Arizona essentially everything it had sought. The decision affirmed the lower-basin allocations authorized by Congress in 1928—4.4 million acre-feet for California, 2.8 million acre-feet for Arizona, and 300,000 acre-feet for Nevada—and ruled that diversions from Colorado River tributaries within Arizona, California, and Nevada should not be counted against those states' allocations, a provision that really affected only Arizona. The decision also granted the secretary of the interior the ultimate authority over the river, with the power to ignore the doctrine of prior appropriation and, if necessary, to overrule existing agreements between states.

Perhaps the most important result, from Arizona's point of view, was that the affirmation of its mainstem allocation meant that it could now pursue its own major diversion. Securing congressional approval and federal funding took five more years and a huge amount of political deal-making. Construction didn't begin until five years after that, and getting water all the way to Tucson took another twenty years. But the first big impediment to the Central Arizona Project was out of the way.

· · ·

GRADY GAMMAGE, JR., a lawyer and real estate developer in Phoenix, is a former president of the Central Arizona Project's board of trustees. His name is a familiar one in the Phoenix area: an auditorium at Arizona State University is named for his father, who was the university's president from 1932 until 1959, and he himself has been active in civic affairs for a long time. We met in his office, in a big building in downtown Phoenix. "Arizona from the very beginning has had a chip on its shoulder about this river," he said. "We felt that, because the Colorado River flows through Arizona for three-hundred-plus miles before it ever forms a border with California, and because there are tributaries of the river that flow out of Arizona into the Colorado, we should get at least as much water as California gets. Now, of course, the reality is that the Colorado River tributaries flowing out of Arizona are pretty worthless—there's not a whole lot of water that gets all the way there. But California's position was essentially, 'Hey, we're California, shut up, we've been using it for a long time, we have lots of Congressmen, we're important and you're not.' And that stuck in our craw for generations."

California tried hard to kill CAP, which it perceived as a serious threat to its economy. It didn't succeed, but its congressional delegation did exact a major concession: a provision stating that Arizona's entire Colorado River water right was junior to California's. Theoretically, that means that during a shortage California would be entitled to divert all of its 4.4 million acre-feet before Arizona took any. Byron E. Pearson, a professor of history at West Texas A&M University, told me in an e-mail that this "California guarantee" had been devised by Northcutt Ely, whom he described as "California's devious but brilliant water strategist." Ely had been a deputy in the Department of the Interior during the presidency of Herbert Hoover, and he was the author of a number of the legal documents that form the theoretical foundation of the Law of the River. He was also a master of the tactical courtroom delay. (Marc Reisner wrote in *Cadillac Desert* that "one expert witness complained that Ely spent three days cross-examining him about a matter that could have been settled in a minute and a half.") Pearson continued, "Essentially, Ely preserved California's claims under the doctrine of prior appropriation within the lower basin's compact allotments—exactly what the compact of 1922 was intended to negate."

This is all true, although it's impossible to imagine the federal government allowing California to dry up Arizona. When Governor Moeur was fighting Parker Dam, he argued that once California had become dependent on surplus water from the river—which the terms of the compact entitled it to use—Arizona would never be able to take its full share, because doing so would threaten the continued existence of what by then would be established cities and farms in California. He had a very good point, and much the same argument would apply today, basin-wide. It is inconceivable that the Department of the Interior would allow Los Angeles and the Imperial Valley to fully cut off water to Phoenix and Tucson (combined metropolitan population: 5.5 million) simply because a fifty-year-old congressional bargain made California's water right senior. But in a serious shortage Arizona would suffer first, and that fact is still a source of statewide aggravation.

The Central Arizona Project is the largest, most expensive aqueduct system ever built in the United States. It begins at the southern end of Lake Havasu, at the Mark Wilmer Pumping Plant, which is yet another piece of western water infrastructure named for a lawyer. (It was Wilmer who made the dramatic and ultimately successful shift in the state's legal strategy during its 1952 Supreme Court case; his opposing counsel was Northcutt Ely.) The system includes fourteen pumping plants, which consume roughly a quarter of the electricity produced by the coal-burning Navajo Generating Station on the southern shore of

Lake Powell. An environmentalist told me that the energy consumed by CAP is roughly the amount that would be required to bring the same volume of water to a boil. The pumping plants lift the water a total of nearly three thousand feet and propel it through fifteen miles of tunnels, ten enormous pre-cast concrete siphons, and roughly three hundred miles of open canals, and en route they temporarily park some of the water in Lake Pleasant, a major reservoir on the Agua Fria River. The canal sections are lined with concrete and are eighty feet wide at the top, twenty-four feet wide at the bottom, and sixteen and a half feet deep. The total construction cost was at least $4 billion—much more than anyone had expected. There was talk, during the planning stages, of covering the canal to reduce evaporation in the desert, but doing so would have hugely increased the cost without saving much water. The system today loses less than five percent of its volume to evaporation each year, and three-quarters of that loss comes from Lake Pleasant, which couldn't have been covered anyway.

CAP was financed by the federal government, although Arizona is required to pay much of the money back, eventually. (The federal government itself covered the cost of elements that are considered federal obligations, such as the dam that created Lake Pleasant and all parts of the project that are involved in supplying water to Indian tribes, and by law Arizona doesn't have to pay interest on CAP water used for agriculture.) Arizona's original plan for reimbursing the Treasury depended heavily on selling CAP water to farmers, who at the time were irrigating mostly with groundwater. The farmers supported that plan when it was proposed, but, by the time the project was fully functioning, circumstances in the state had changed and they no longer wanted the water. Gammage told me, "The first wrong assumption was that the canal was going to be cheap, and it turned out to be really expensive. The second wrong assumption was that Arizona would continue to be the premier source of the world's finest cotton—Pima cotton—which would carry a market price that would enable the farmers to pay for all the water they could take. Well, by the time the canal was finished, Pakistan, India, and Egypt were all growing high-quality cotton, and the market had become much more globalized. Cotton farmers in Arizona were no longer making a lot of money, and they couldn't afford what the fully loaded cost of the canal was going to be—and yet if we didn't buy the water and use it California could take it all. Keeping water away from California is one of our fundamental principles, so we had a crisis."

Gammage's first exposure to western water management occurred when he was in high school, in the late 1960s. He got a summer job with the Salt River Project, one of the country's oldest reclamation efforts, which operates four reservoirs in Arizona and supplies water and electricity to much of Phoenix. (A friend of mine lives in a part of greater Phoenix that began as a Mormon settlement in the 1870s. The community has extremely old water rights, and the yards there— including my friend's, which covers three acres—are flood-irrigated with SRP water, in his case at a cost to him of four dollars an acre-foot.) In 2007, in an interview that was part of an oral history of CAP, Gammage explained that the economics of water in Arizona in the sixties were much different from what they became later. Among his tasks one summer, he said, was making an inventory of vehicles owned by the project. "We actually discovered that they had bought a fleet of trucks several years before, parked them on a piece of property, and forgot they bought them. This was in the really plush days."

The plush days were over by the early nineties. To solve its CAP funding problem, Gammage told me, Arizona had to find a way to make the Colorado River water cheap enough for farmers to take it and use it in preference to groundwater. "We ended up making a deal with the farmers that was suggested by a task force appointed by Governor Symington," he continued. "We told the farmers that we would sell them Colorado River water at significantly less than the cost of getting it to them—and I mean significantly—and that the cities would cover the gap." In effect, Phoenix, Tucson, and other municipalities agreed to take CAP water they didn't need, then sell it to the farmers at a deep discount. "The cities were willing to pay," he said, "because they were afraid that, if we didn't get the canal fully flowing and operating, over time California's usage would become an irrevocable entitlement." And if that happened, the state's urban planners believed, the growth of Arizona's cities and their economies would end.

One of CAP's purposes was to bring Colorado River water to Tucson, which used to be one of the biggest cities in the world that relied exclusively on groundwater. Tucson sits above a large aquifer, which is fed mainly by precipitation in the mountains that surround it—and, in fact, the name "Tucson" comes from Indian words meaning "water at the foot of black mountains." As the city grew, however, it drew down the aquifer faster than percolating snowmelt and rainfall could replenish it. The Tucson water department had to keep drilling deeper, and, as it did, the quality of the extracted water fell and the cost of lifting it to the surface rose. Nevertheless, the transition to CAP was at first a disaster. Water from the river had a different chemical composition and contained more solids than water from the city's wells, and it was pumped through the municipal pipe network at a different pressure and in the opposite direction—and those factors, in combination, loosened many years' worth of accumulated crud inside the pipes, creating disastrous leaks and turning everyone's water brown. The immediate result, Gammage said in 2007, was that "the people of Tucson immediately rose up in an outcry that this was a plot by Phoenix to poison them." Tucson eventually got better at treating, blending, and pumping CAP water, and, after various referendums and high-level firings, the city won back the support of its water customers. One of the key steps was a campaign in which the water department gave away bottles of new and improved CAP water to demonstrate that it could be both transparent and nontoxic.

Long before that water got to Tucson, President Carter tried to kill CAP, along with dozens of other federal water projects. That effort accomplished virtually nothing, other than ruining whatever chance Carter had of building a working relationship with Congress. Marc Reisner, almost a decade later, wrote, "To a degree that is impossible for most people to fathom, water projects are the grease gun that lubricates the nation's legislative machinery." The currency of congressional pork-swapping has changed since then, partly because of public reaction to revelations that Reisner himself made in *Cadillac Desert*. But water projects were hugely important at the time, and even people who eventually became opponents of all kinds of water-related boondoggles were angry about Carter's dam-and-reservoir "hit list." Morris Udall— Bradley Udall's father—was an Arizona congressman from 1961 to 1991, and was among the people who had been instrumental in getting CAP approved in the first place. In the later years of his career, he regretted a number of the water projects he had helped to push through Congress, including that one, and was prominent in environmentalist causes. But in 1977 he said that, if Carter succeeded in killing CAP, Tucson and Phoenix would "dry up and blow away." Carter's efforts on water are generally viewed today as having been ahead of their time, in terms of federal action on a major environmental issue, yet also extraordinarily ill-conceived and poorly managed. And they didn't succeed in killing CAP.

12.

THE RULE OF CAPTURE

My wife and I bought our house in northwestern Connecticut in 1985, and among our first acquisitions after we took possession were a hose and a lawn sprinkler, which I set up in a patch of dry grass near the back door. We had lived in an apartment in Manhattan until then, and I simply assumed that one of my responsibilities as a homeowner was to make my yard green.

An hour or two after I began watering, though, I noticed that my new sprinkler had stopped sprinkling. I figured the hose must have become kinked—but, no, the hose was fine. What had happened, I eventually realized, was that our water had run out. That was the first time I truly understood that the water in our new house came from a well in our yard, and that the water in the well came from somewhere underground. (My initial reaction was a feeling of tremendous relief: I'll never have to water my lawn again. And I never have.) We still live in the same house, and during the past three decades I've thought about our dependence on groundwater only intermittently: a few times when lightning has fried our well pump, once when a contractor crashed his backhoe into the well casing, again when a plumber explained that our water was so acidic that it was eating our plumbing, and on various occasions when snowstorms have knocked out our electricity, causing our well pump to stop pumping and, therefore, our toilets to stop flushing. Most of the rest of the time, though, like most Americans, I simply assume that when I turn on a faucet or a shower water will come out.

The town my wife and I live in has a small private water company, which is more than a century old and serves fewer than two hundred customers. Our closest neighbor is connected to it, but our house is just beyond the service area. Some years ago, the system suffered a major line break in the middle of our village center. A friend of mine, who was the company's (pro bono) attorney, noticed commotion from his office window and walked across the street to see what was going on. He helped restore service, by running a garden hose across the driveway of the gas station to a spigot at the grocery store. He told me that, at a rate-increase hearing once, a state commissioner had asked him to describe the company's emergency systems, and he said, "When Mrs. Eames can't get water in her upstairs bathroom, we know the reservoir is low." (Only the court reporter laughed, my friend said.) And the "reservoir," in those days, was actually just a spring-fed puddle, with a rotting wooden roof over it. The company's water-treatment protocol consisted of hiring a local man to remove drowned mice with a net. (He was paid by the carcass.) Today, the water company still taps the same groundwater sources that fed the old springs, but it does so with wells.

Most of the residents of my town get their water from wells on their own property, as I do. Drilling a well around here is a little like buying a lottery ticket. The one in my yard is a little over 270 feet deep and produces something like three gallons a minute. A friend who lives a third of a mile away has a 150-year-old dug well, which is twenty-two feet deep and, except during very dry summers, produces more. Another friend, a few miles in another direction, has two wells, both very deep, which between them produce less than two gallons a minute. Yet another friend, about the same distance away, gets more than sixty gallons a minute. Earlier this year, the owners of an enormous house not far from my house drilled a new well, and added a buried ten-thousand-gallon storage tank, to feed the irrigation system they had just installed to water their enormous yard. And a few months later one of their neighbors (who is served by the private water company but had just added an even bigger irrigation system to their even more enormous yard) dug a well, too—and then added a second one.

All of these people, I'm pretty sure, view the water they pump from their wells in the same way I have always viewed mine. That is to say, they think of the water under their property in the same way they think of the grass on top of it, as something that belongs to them. There's a creek at the bottom of the hill our house stands on. It doesn't touch our property, but even if it ran directly past our door I would have no difficulty viewing it as a public resource—as a common good, to be shared with others. But the water directly under my yard, the water I pump from my own well: that's mine.

WHEN THE WESTERN STATES were sorting out their ideas about water rights, in the late nineteenth and early twentieth centuries, most people thought of groundwater in the same way. They thought of it as a form of personal property, and they also thought of it as entirely distinct from surface water. And in many places today the same distinction persists in law. (If I ever wanted to drill a new well, or to make my existing well deeper, I wouldn't expect to have any trouble securing a permit—application fee: twenty-five dollars.) One of the oldest guiding principles governing groundwater extraction is the so-called Rule of Capture, a legal doctrine that grants landowners the right to use as much water as they want to from wells on their own land, even if doing so dries up wells belonging to their neighbors. In his book *Water Follies*, published in 2002, Robert Glennon, a professor at the law school of the University of Arizona, describes a 1904 case in which the Texas Supreme Court "refused to protect adjoining landowners because it thought that the principles that control the movement of groundwater were 'so secret, occult, and concealed that an attempt to administer any set of legal rules [would result] in hopeless uncertainty, and would, therefore, be practically impossible.'" There is no mention of groundwater in the Colorado River Compact.

Nowadays, the fact that a hydrologic connection almost always exists between water on the surface and water underground, except in places where the two are separated by an impermeable stratum, is neither "secret" nor "occult." Buzz Thompson of Stanford Law School told me, "In the West, if you find a stream that flows year-round, it's generally because there's groundwater feeding it"—and that relationship is reciprocal, since diverting water from surface streams can also push down water tables. Every well creates a "cone of depression" in the subterranean material it draws water from, and if the extraction rate is high enough the cone can extend over a great distance. In coastal regions especially, the consequences can be dire. The largest single source of freshwater in a large portion of the southeastern United States is the Floridan Aquifer, which underlies all of Florida and parts of Alabama, Georgia, Mississippi, and South Carolina. In 2012, an official of the Florida Geological Survey told me, "We've had well fields that were pumping groundwater from close to the coastline, and those well fields eventually began pulling in saline water. There's an interface between freshwater and seawater, and as those wells were moved inland they kind of dragged the interface with them." During the past fifteen years, the water utility that serves Hilton Head, South Carolina—which lies near the northern edge of the Floridan—has had to abandon more than half a dozen ruined wells, as saltwater intrusion has spread. This is a global issue, and it's an increasingly important one because groundwater in coastal areas is directly affected by changing sea levels: as oceans rise, they push harder against freshwater interfaces. The effects can extend over large distances. The big earthquake in Japan in 2013 made water elevations in wells in Florida fluctuate by three inches.

Still, old perceptions linger, in the law and elsewhere, and the fact that groundwater is usually invisible makes it hard to regulate and easy to overuse. Irrigated agriculture in the Great Plains depends heavily on water from the Ogallala Aquifer, which underlies parts of eight western states, including three Colorado River Compact states (although in those three states the aquifer is well outside the Colorado's drainage basin). In 2014, the USGS estimated that withdrawals from the Ogallala since pumping began in earnest—essentially, over the past fifty to seventy-five years—have totaled more than 300 million acre-feet, or roughly twenty-five times the current volume of Lake Mead, and that withdrawals between 2011 and 2013 alone amounted to 36 million acre-feet, or roughly three times the current volume of Lake Mead. In some agricultural areas of the Great Plains, the water table has fallen so far that pumping is no longer economical. And in many cases once Ogallala water is gone it's gone. Most of it is fossil water, which entered the ground no more recently than the last ice age. There's so little precipitation in the overlying areas that most parts of the aquifer recharge extremely slowly—too slowly to make a difference to anyone alive today.

In recent years, a group of scientists has attempted to quantify groundwater depletion all over the world by using data from a NASA mission, the Gravity Recovery and Climate Experiment. GRACE employs two satellites, which follow each other in orbit around the earth. James S. Famiglietti, one of the scientists, told me, "Each of the satellites is about the size of a squashed minivan. They're separated by two hundred kilometers, and the primary thing they measure, using infrared lasers, is tiny changes in that distance, plus ups and downs." The fluctuations are caused by variations in the earth's gravitational pull, which are caused by variations in the mass of whatever the satellites are passing over. "It works a little like a scale," he continued. "When you step on a scale, gravity is pulling you toward the earth, and the heavier you are the farther it pulls. Same principle." The mission's original focus was on such obviously climate-related phenomena as melting glaciers and rising sea levels, but the scientists eventually realized they could detect changes in groundwater volumes, too.

"It took a while for the picture to come into focus," he said. "The first papers were about ice sheets, about Greenland, about Antarctica, and no one was really looking at the continents yet." For him and his colleagues, he said, the "holy crap moment" came while they were studying data from India. "We saw this big hot spot and investigated it, and realized that what it showed was groundwater depletion. And then, sure enough, you start looking around the global map and you see all these other hot spots, and you know from the geography that they are sitting on top of all the major aquifers of the world." Among the hot spots they noticed were large ones in the western United States.

The GRACE satellites, even though they measure micron-scale fluctuations in inter-satellite spacing, are not precision instruments when it comes to monitoring groundwater. They draw their data from large terrestrial areas, and because they measure mass in general rather than groundwater in particular their results have to be adjusted to account for rainfall, snowfall, storage levels in reservoirs, soil moisture, and other variable conditions—the equivalent of subtracting the weight of your shoes and clothes from the reading on the dial of your scale, except that, when it comes to water, it's much harder to tell where the shoes and clothes stop and the person on the scale begins. Still, Famiglietti told me, he and his colleagues are confident that they know what they're looking at. "Globally," he said, "the worst danger spots are India, Bangladesh, Pakistan, the Arabian Peninsula."

More than a few of the regions that are suffering the gravest impacts from exhausted aquifers are the same regions that gained the most from the Green Revolution, which, after its start in Mexico in the 1940s, transformed agriculture in some of the poorest, driest parts of the world by drilling wells, improving irrigation systems, extending access to synthetic fertilizers and pesticides, and introducing high-yielding varieties of staple grains (most of which depend heavily on irrigation, synthetic fertilizers, and pesticides). Those advances transformed Third World food production, and turned some poor countries from reservoirs of malnutrition into net food exporters. But they also helped to create what have turned out to be unendurable environmental pressures, by draining aquifers, exhausting soil, and allowing populations to soar to levels that can't be sustained.

Groundwater depletion affects wealthy countries, too. Beginning in the 1980s, Saudi Arabia became a serious producer and then a major net exporter of wheat. It did that by tapping an aquifer that was estimated to hold as much water as Lake Erie, and using enormous center-pivot irrigation systems to turn much of the overlying desert into farmland. But the water was almost entirely fossil water, and, after more than two decades of hard pumping, the aquifer is now essentially empty. The Saudi government began shrinking the irrigated area in 2008 to slow the rate of depletion, and in 2016 it permanently abandoned it, ending its twenty-five-year experiment in locavorism. This is not an issue of "sustainability"—the amount of water in the aquifer was finite, so there was no permanently supportable rate of withdrawal—but it's a good example of the economic and geopolitical impacts that resource exhaustion can have, and over very short periods of time. In less than a decade, Saudi Arabia went from a major wheat exporter to a total wheat importer, a swing of many billions of dollars. The Saudis have addressed that change in part by using oil profits both to import food and to purchase agricultural land in other parts of the world, including North America.

"Even worse than the Arabian Peninsula," Famiglietti continued, "is what I call the Northern Tier: Turkey, Syria, Iraq, Iran. Those four are an order of magnitude worse than, say, Israel, Jordan, Palestine, Saudi Arabia, Yemen—which are terrible, don't get me wrong, but the Northern Tier is worse." Syria is an especially unnerving example, since the ongoing civil war there was closely preceded, and was almost certainly at least partly caused, by catastrophic aquifer depletion accompanied by record-breaking drought, both of which forced large numbers of unemployed men to migrate from ruined farms to crowded cities. (The GRACE team has estimated that, between 2003 and 2009, the Northern Tier ran through 117 million acre-feet of stored freshwater.) Indeed, you could make a plausible argument that most wars, if not all of them, are at least partly resource conflicts—both in cases where powerful countries or factions attack weaker ones in order to gain or protect access to economically important raw materials, and in cases where weaker countries or factions attack more powerful ones in response to resource exploitation. The ongoing migration crisis in Europe may be a foretaste of crises to come.

The GRACE program is a crude tool for measuring water on land, and the scientists who gather the data are still trying to figure out how to interpret it. But the satellites help to fill in the big picture, and they reinforce the obvious lesson that you can't suck water from the ground forever without creating problems above the surface. In this country, Famiglietti told me, the most depleted areas include California's Central Valley, the southern high plains, "all across the southern part of the United States," and the Colorado River basin.

"THE STEADY DRAIN on underground reserves grows out of two realities," Robert Glennon writes in *Water Follies*: "Canals and pipelines don't reach far enough to deliver surface water to everyone, and laws don't reach far enough to stop people from drilling." Among the Colorado River states, the one with the most primitive approach to groundwater is probably California, which, even before the drought, was getting most of its water from wells. In Los Angeles in the late 1800s, there was so much water in the ground that artesian wells sometimes spewed it into the air. That era ended rapidly. There are places in the San Joaquin Valley where the surface of the land has subsided by dozens of feet because groundwater there was pumped so much faster than the environment could replenish it. The San Joaquin Valley is well outside the area served by the Colorado, but the fates of the two regions are linked because any major decrease in California's ability to draw water out of the ground necessarily increases its reliance on its compact allotment—and that makes cutbacks in Colorado River diversions both more consequential and harder to achieve. And that relationship works in the other direction, too, of course, since cutbacks aboveground make water underground both more tantalizing and more valuable. Even so, California has only recently begun to talk seriously about controlling, or even measuring, groundwater withdrawals.

Colorado's legal handling of groundwater is more sophisticated. Since the 1960s, it has classified almost all its water, both on the surface and underground, as "tributary water"—a category that includes not just all streams and rivers but also all groundwater that's hydrologically connected to them. Extractions of tributary groundwater are governed by the water courts in essentially the same way that diversions from rivers are, and wells that tap tributary aquifers have priority dates and water-court decrees in essentially the same way that ditch systems do. That means that a well can call out a surface diversion if the well's right is senior to the ditch system's and there isn't enough water in the combined system to satisfy both.

Colorado water law makes exceptions of various kinds for four types of groundwater: "exempt wells," "designated groundwater," "non-tributary groundwater," and "not non-tributary groundwater." To an English major, at least, "not non-tributary groundwater" and "tributary groundwater" sound as though they must be the same thing, but they're not, exactly. "Not non-tributary groundwater" is tributary groundwater that, for practical and other reasons, is treated legally as though it were non-tributary. The category was created by the state legislature mainly to accommodate residents of the Denver Basin, who have little choice but to draw down and, eventually, exhaust certain aquifers that are hydrologically connected to surface streams—because if they couldn't do that they'd have to import even more water from the other side of the mountains than they do already. And there are plenty of other complicated provisions. For example, non-exempt wells with junior priority dates don't have to shut down when surface flows are low but do have to have an approved "augmentation plan," by which their owners agree to take steps (or pay money) to provide replacement water to senior rights holders who would otherwise be shorted when there isn't enough water to satisfy both.

Arizona's rules concerning groundwater are at least as confusing as Colorado's. In 1980—in an effort both to resolve increasingly acrimonious conflicts among water users and to assure federal approval of the completion of the Central Arizona Project—the Arizona legislature passed the Groundwater Management Act, a hugely ambitious bill. Its purpose was to exert control over groundwater withdrawals before the state's most important aquifers had been depleted. The most stringent requirements apply to regions that the act designated Active Management Areas—within which, for example, land sales are forbidden unless the developer is able to demonstrate to the state that the proposed development has an "assured water supply" that is sufficient to last it for a hundred years. But doing this is so easy that it isn't much of an impediment.

Other provisions of the Groundwater Management Act, and of a subsequent piece of legislation, the Groundwater Replenishment Act, are concerned with refilling aquifers. Even once the CAP infrastructure had been completed, Arizona wasn't able to use all the water it was entitled to divert under the Colorado River Compact. To keep California from taking that water, Arizona began diverting it anyway and storing it underground, by flooding it onto "spreading basins" in the desert and allowing it to soak in—a practice known as "water banking." The advantages of banking water go beyond interstate rivalry, because water stored in aquifers, unlike water stored in reservoirs, doesn't evaporate. (Arizona, during the past decade, has banked millions of acre-feet in this way.) The Groundwater Replenishment Act also enabled real estate developers to satisfy their hundred-year-assured-supply requirement by banking water in other places—and further refinements enabled them to do so without actually putting wet water in the ground. For example, when Arizona farmers irrigate with CAP water, the state now classifies the groundwater they didn't use as though it were new water that had been banked—a semi-fictional accounting maneuver known as "indirect recharge." Most of the legally mandated groundwater replenishment that takes place in Arizona is done that way. Gammage characterized the practice to me as "awesome—a little scary, but interesting."

Some of the most innovative water-banking practices in the West originated in Nevada, with Patricia Mulroy. In the late 1980s, the Southern Nevada Water Authority began treating unused portions of Nevada's Colorado River allotment to potable standards and pumping it into dedicated wells near Las Vegas. In 2014, Andrew Burns of the SNWA told me, "To date, we've put about 365,000 acre-feet of treated water into the ground. Doing that has had several benefits. One is that it's a future supply, which we can draw on when we need it. Another is that it has helped to maintain our aquifer, by raising the water table by about 120 feet." Southern Nevada depends mostly on the Colorado River for its water, but it still gets about ten percent from the ground—

THE RULE OF CAPTURE

a withdrawal rate it has been able to sustain without lowering the water table.

In recent years, as Lake Mead has continued to shrink, Nevada has left some of its compact allotment in the lake and "stored" the unused portion not in the ground but in California. It doesn't actually ship the water across the state lines; instead, California agrees to keep the equivalent amount somewhere in its reservoir system, in effect with Nevada's name on it. "At the moment, we've got about 150,000 acre-feet stored with them," an SNWA official told me in 2014. "When we need it back, we'll take that amount from Lake Mead and they'll forego a portion of their Colorado River allocation." Nevada pays Arizona to do something similar, using both direct and indirect groundwater recharge. "They pay us to bank it," Gammage told me, "and then at some point in the future we can refrain from taking some of our own water out of the Colorado, and they can take the same amount from Lake Mead—and we can charge them again. It's a way of getting gaming revenue in Arizona without gaming. And it makes Nevada an ally in dealing with California—which is very valuable."

He continued, "When I talk to lay groups about water, people always ask, 'Do we have enough? Are we going to run out?' And I always say that that's the equivalent of my asking whether you have enough money. Because the answer depends on what assumptions you make and how much risk you're willing to take. In Arizona, we've gone from an era when groundwater was cheap, abundant, free, and inexhaustible— which is how they still treat it in places like Kansas and Texas—to an era when, because of the Groundwater Management Act, the presumption is that groundwater should never be touched. To the extent that the average person thinks about this at all, they tend to feel that groundwater should not be used, because it's what you have in savings to protect against crisis. But now there are really two different kinds of groundwater in the Active Management Areas. One is this artificially created groundwater, whether indirect or direct recharge, which is like your retirement account. And the other is genuine old-style prehistoric groundwater, which is like what you inherited. Do you want to burn through that, or pass it on to your kids? And then there's surface water—that's your cash flow. Do you want to use it to have a swimming pool? To support lush landscaping in the desert? To grow crops— many of which may be exported? Or do you want to manage your usage much more carefully, and dial everything back?"

13.

BOONDOCKING

South of Parker Dam, the Colorado winds for about ninety miles through or alongside the reservation of the Colorado River Indian Tribes, which covers nearly 300,000 acres in southwestern Arizona and southeastern California. The reservation was established in 1865, the year the Civil War ended. Its residents include four thousand Mohave, Chemehuevi, Hopi, and Navajo Indians, known collectively as the Colorado River Indian Tribes. As a result of various court cases and legal settlements, the tribes own the right to use more than 700,000 acrefeet of Colorado River water—about a quarter of Arizona's entire allotment. That water is used to irrigate roughly eighty thousand acres of tribal land, much of which is leased to non-Indian farmers.

The second largest town inside the reservation, after Parker, is Poston. It consists of little more than a few dozen houses and a depressing-looking post office, but during World War II it was the site of one of the country's largest Japanese American internment facilities, the Poston War Relocation Center. The center was divided into three separate detention camps, and because of their location, in what was then an especially remote part of the desert, there were no guard towers and the perimeter fence didn't go all the way around: like a dare. In 1999, the National Park Service published an illustrated history of all the country's internment facilities, *Confinement and Ethnicity: An Overview of World War II Japanese American Relocation Sites.* The book says that summer temperatures at Poston were so high, and winds from the desert so strong, that internees named the three camps Roasten, Toasten, and Dustin, and that the lumber used for the buildings shrank so much, because of the dryness and the heat, that the contractor—Del Webb, who three years later became a co-owner of the New York Yankees—had to fill the gaps with millions of feet of wood strips.

Just beyond the southern end of the reservation, on the California side of the river, I looped around to the north and drove up Colorado River Road. Almost all the houses on both sides were trailers or double-wides, and the ones on the right-hand side stood on lots that backed up to the river and had their own docks. RVs and boats were parked in many of the driveways, and some of the lots were surrounded by chain-link security fences with sliding gates. Next to one of the driveways was an enormous saguaro cactus and a cross made from four-by-fours and covered with what looked like plastic rings. The cactus was more than twenty feet tall, and one of its many arms, up near the top, looked like a bunched-up mutant fist. Next to another driveway was a mailbox in the shape of a reduced-scale but still quite large Jet Ski. Its owner was standing next to it, retrieving his mail. As I drove by, I shouted, "Nice mailbox!" and he nodded the way you would if you had just won an award and wanted people to know it hadn't gone to your head. In another driveway, a guy was cleaning a set of golf clubs.

The northern end of Colorado River Road merges into the entrance of Mayflower Park, an RV campground and recreational area owned by Riverside County. I told the guard at the gate that I was just looking, and he let me drive in without charging me for a day pass. "The clear blue of the river mimics the beautiful blue skyline with spacious green grass nestled in between," the website says. The park has 179 RV sites, most of them with hookups, and it has picnic areas, horseshoe courts, shuffleboard courts, a lawn-bowling pitch, places to fish, and a big boat ramp leading down to the river, which at that point is a couple of hundred yards wide. I parked near a picnic area and walked down to the river. The water was low, and it barely seemed to be moving, despite a sign that said swimming was forbidden because of dangerous currents. The long view was nice, though: the beautiful blue skyline mentioned on the website, plus a dog pile of cumulus clouds above the full length of the Dome Rock mountain range, ten miles to the east.

Parking an RV at Mayflower in a site with a full hookup works out to somewhere between $200 and $300 a week, including charges for things like trash collection and holding-tank dumping. At one of the occupied sites, two older men were trying to reposition a satellite-television dish, which had blown over during the night, while their wives sat in lawn chairs and paid no attention to them. One of the men was somewhat taller, balder, and heavier than the other, and one of the wives was holding an aluminum cane and a well-behaved Australian terrier. Halfway between the men and the wives was a metal picnic table covered with a blue-and-white-checked plastic tablecloth, and on top of the tablecloth were a propane cooktop, a Culligan water filter still in the box, a hose nozzle, a foot-tall green plastic Christmas tree attached to a power strip, some Bungee cords, a toaster-size glass terrarium with a gift bow tied around it, and a large ceramic smiling frog sitting on a ceramic log embossed with the word "Welcome." I introduced myself and tried to help with the satellite dish, but mostly I got in the way. The dish's owner told me that he and his wife were from Oregon, and that the other couple, whom they had just met, were from Canada. "A lot of the people here are from Canada," he said. "If you stay for any length of time, I highly recommend a water softener. The water here is like liquid concrete." That water comes from a well, not the river, and Mayflower Park had been having serious problems with it for a while. At the other end of Colorado River Road, on a peninsula-like bend in the river, I drove through a larger, family-owned campground, Hidden Beaches Resort. Most of the people spending the winter there were doing it not in RVs but in mobile homes that looked like permanent installations. An online reviewer a few years ago complained that the river at Hidden Beaches looked "drained and muddy."

Mayflower Park and Hidden Beaches are on the outskirts of Blythe, which is the biggest town in the Palo Verde Irrigation District, a long, skinny agricultural area that borders the Colorado on the California side and draws water from it. Palo Verde's water right has a priority date of 1877. The intake is at a thousand-foot-wide diversion dam and spillway at the northern tip of the district, five or six miles upstream from Mayflower Park. From the dam, the water flows into a "desilting basin," and then through a network of canals and ditches to roughly ninety thousand acres of cultivated land. I drove past a recently plowed field whose furrows were so straight and sharp and smooth that the field looked like brown corduroy pulled tight on an artist's canvas stretcher. The soil was as dark as chocolate, and for a moment I thought it must have been brought in from somewhere else—maybe India, where real estate developers in the United Arab Emirates buy topsoil for golf courses—but of course it was just silt deposited by the wandering Colorado. I saw fields in every condition from recently harvested to about-to-be harvested to recently planted to empty. I also saw warehouse-size stacks of baled forage crops, mainly alfalfa and Sudan grass. And I stopped briefly at the main entrance of the main section of the Palo Verde Ecological Reserve, a thirteen-hundred-acre conservation area that runs along the river for a few miles, beginning a short distance upstream from Mayflower Park. The reserve was created in 2007 by the state Fish and Game Commission plus various partners. It consists almost entirely of former farmland, which has gradually been planted with cottonwood, mesquite, and other native riparian vegetation. The project is an attempt to re-create something like a natural floodplain habitat, and the old irrigation levees that crisscross the replanted fields make good hiking and biking trails. Hunting is allowed; a sign near the parking area said "Shotgun Shells Are Litter, Too!"

In 2003, state and municipal water authorities in California, together with the U.S. Department of the Interior, negotiated the Quantification Settlement Agreement for the Colorado River, a pact that is being phased in and will take full effect in 2021. Its purpose is to reduce California's reliance on the Colorado River to its compact allotment, 4.4 million acre-feet, and to "transfer" a significant fraction of that water from farms to Los Angeles and San Diego. In 2004, in conjunction with the agreement, the Palo Verde Irrigation District made a long-term deal with the Metropolitan Water District of Southern California to annually supply municipal users there with as much as 118,000 acre-feet by fallowing, on a rotating basis, up to twenty-eight percent of the district's farmland. As I drove around, I saw many fallowed fields—which stand out from cultivated fields, even ones on which crops have been planted but haven't begun to come up yet, because they look as though they were quietly trying to turn back into desert. Landowners receive a bonus payment for signing up for the program, then an additional fee for every acre they take out of production. They're allowed to apply small amounts of water periodically to keep their fields viable and their irrigation systems functioning, and fields that have been fallowed for a year are typically more productive when they're planted again, after "resting." The 2004 agreement runs for thirty-five years and could eventually transfer a total of almost four million acre-feet of Colorado River water from the district to metropolitan Los Angeles.

· · ·

FROM BLYTHE, I headed east across the river to Quartzsite, Arizona. It's famous partly for being one of the hottest places in the United States, partly for being what the town's website calls the Rock Capital of the World, and partly for being what may be the country's largest wintertime RV assembly point. The town has a permanent population of only about thirty-five hundred, but it attracts more than two million visitors every year, most of them during months when my yard in New England is covered with snow. Almost all the visitors arrive in RVs or in trucks pulling some sort of trailer. (There's only one motel, a Super 8, within twenty miles of the town center.) The area doesn't have enough wired and plumbed campsites for all the people who show up, so the majority "boondock," by parking in the desert on one side or another of the highways that intersect there, mostly on Bureau of Land Management land. Sam Penny, an RVer and a travel blogger, first visited with his wife, Alice, in 1999 and wrote, "I decided the main thing when boondocking at Quartzsite was to park in a place that had intrinsic beauty. Later I found that one should also look for some place where those with generators would not gather, at least if you happen to be a solar boondocker like we were."

The World Wide Web and Amazon.com have been a great gift for older people who now live mainly in RVs and have things on their mind. The most recent posting I could find on any of Penny's numerous blogs is from late 2014, so I fear the worst, but during his traveling and publishing heyday his principal areas of interest included the trips he took with Alice; outfitting his RV, a fifth wheel with slide-outs; the Ebola virus; solar power and LEDs; the possibility of a new killer earthquake occurring on the New Madrid Fault in the Mississippi River Valley (a disaster he explored in two science-fiction novels); and "the Great Collapse of our civilization." Just south of Quartzsite, I passed a huge motor coach parked by itself in an empty expanse of desert next to the highway. The owner had positioned his vehicle so that one of its long sides faced the road, and he had hung up an enormous, not-inexpensive-looking banner announcing that books he had written were available online (and also, I assumed, in the motor coach). I wanted to drop by, but I couldn't figure out how to get to the place where he was parked without registering at a BLM campsite.

I know something about RVs. My father rented a twenty-seven-foot Dodge motor home in Denver in 1967 after picking me up at summer camp in Florissant, and then he and my mother and sister and brother and I spent a week traveling around Colorado in it. My father had trouble getting it up Monarch Pass in a storm and then much more trouble getting it down on the other side, and one night the holding tank suddenly overflowed all over the floor, and sleeping with the generator running turned out to be even harder than sleeping with the generator (and therefore the air conditioner) turned off. My mother assumed that these and other hardships would kill my father's interest in RVing, whose sudden emergence had taken her by surprise, but instead they inflamed it, and during the next eighteen years he owned three. Each new one was larger and more luxurious than its predecessor, and each was known in our family and to my parents' friends as the Bus. When I was in high school, I was allowed to borrow the first two Buses. The look on the face of a mother whose daughter is being picked up to be taken to a drive-in movie by a high school junior driving a Greyhound-size motor home is impossible to convey with words. We first got a color TV in our house in the early seventies, because the second Bus had come with one and my mother complained. The third and final Bus was thirty-five feet long. It was custom-built to my father's exacting specifications by Newell Coach, a company in Oklahoma, and many of its numerous ingenious features were designed around cocktail hour. My father parked the Bus at the end of our driveway on a thick concrete pad that had to be repoured because the first pour (he had deter-mined) wasn't perfectly level.

I myself have rented RVs twice: the first time in 1991 for a *New Yorker* assignment during which I traveled around New Jersey with the performers of a one-ring circus, and the second time for that big west-ern trip my wife and I took with our kids in 2006. Neither experience made me wish I owned an RV, but my wife and I have often talked about renting one again. Her interest is surprising in a way, because she's a reluctant traveler under almost any circumstances, but not sur-prising in another, because there is something deeply comforting and seductive about traveling around in a semi-self-sufficient and somewhat womb-like container and spending time in welcoming communities of fellow adventurers who decorate their vehicles with ceramic frogs and signs that say "If the RV's a-Rockin', Don't Come a-Knockin'." (Sam Penny, describing a campground in one of his blogs, wrote, "At times it can seem like a hotbed from Peyton Place, if you know what I mean.") Campgrounds where people park RVs are fun to walk around in be-cause you never run out of interesting things to look at and think about, as on the sidewalks of Manhattan. Several times during non-RV travels of my own, I've met retired people who told me they had sold the house in which they raised their children and now lived solely in their RV, moving whenever they got bored with wherever they hap-pened to be. I can't imagine doing that myself, but I can imagine what it must be like: perfect liberty in your own covered wagon, minus the horrors of the real Old West.

For people who own RVs and hate what winter is like in the place where they usually live, the southern reaches of the Colorado River exert a strong gravitational pull, which is amplified by the presence of multitudes of other people who feel the same way. In Quartzsite, the main draws include RV and camping shows; rock, gem, and fossil shows; craft shows; antiques shows; and flea markets. The very biggest shows, which attract many thousands of visitors, are held in January and February. I was a month too early for those, but large numbers of winter visitors had begun to arrive, and I did stop at an open-air stand in the center of town where I could have bought Dremel bits, carpeting, Minnesota pipestone, decorative wall hangings, and push brooms. "We saw ground polished rock balls from marble size to bigger than a basketball," Sam Penny wrote in 1999. "I am not sure what one does with rock balls, but some of my friends might know." I could have bought geodes, fake-stone garden gnomes, equipment for turning ordinary-looking rocks into gems, fossilized dinosaur turds, and clocks that looked like other things. Also, if I'd needed it and had something to hold it in, I could have bought freshwater from a freshwater filling station whose storage tanks had been decorated to look like yellow-and-black floppy-eared elephants. (They dispensed water from their trunks.) Water filling stations—some of which look like gas stations and a few of which look as though they might have started out as one-hour-photo kiosks—are common in the entire region, and for people who boondock they're a necessity.

I also visited Quartzsite's cemetery. In the era before RVs (and also before trains, airplanes, and highways), the baking deserts of the southwestern United States posed a serious transportation conundrum: how do you get from one side to the other? One possible solution, first proposed in the 1830s, was camels. Jefferson Davis, who became Franklin Pierce's secretary of war in 1853, was an enthusiast, and in 1856 he persuaded Congress to spend $30,000 on a test. The Army eventually imported about seventy camels, mainly from Northern Africa, the Middle East, and Mongolia, along with a few experienced handlers. One of the handlers was a Syrian (or possibly a Greek, Jordanian, or Turk) named Hadji Ali, who in this country was usually known by an Americanized nickname: Hi Jolly.

In most respects, the experiment worked extraordinarily well. The camels carried heavy loads, and contentedly ate desert vegetation that other Army animals wouldn't, and traveled extremely long distances between water breaks. Hi Jolly once saved five American soldiers pinned down by Indians by racing toward the Indians on a camel while waving a scimitar and shouting in Arabic. But the camels didn't get along with the Army's horses and mules, and most of the soldiers were suspicious. Then the South seceded from the Union, and Jefferson Davis turned his attention elsewhere. Some of the camels were later sold or given away, and some were set loose in the desert, where their descendants were spotted, occasionally, as late as the 1940s. Hi Jolly—who by then was known as Philip Tedro—died in Quartzsite in 1902. In 1935, the Arizona Highway Department erected a pyramid-shaped monument over his grave, and the town later renamed the entire cemetery after him. The monument is made partly from large pieces of petrified wood—chunks of which are always available for sale at the rock, gem, and fossil shows in town. The principal speaker at the monument's dedication ceremony was Governor Moeur, who was taking a break from his water war with California.

DOWNSTREAM FROM the Palo Verde Irrigation District, the Colorado flows through two large national wildlife refuges, Cibola and Imperial, which between them contain more than forty thousand acres of oasis-like wetlands and meanders and backwater lakes. There are no dams, diversions, or farms within the boundaries of the refuges, and there are no RVs except in a couple of parking lots and on viewing roads near the visitors' centers. Imperial includes what its website describes as "the last un-channelized section" of the river before it leaves the United States. Camping is not permitted in either refuge, although certain amounts of fishing and hunting are. To get to the Imperial visitors' center, you drive south from Quartzsite on Highway 95 for sixty miles, then cross part of the Army's Yuma Proving Ground, an enormous military facility, which covers a little more than thirteen hundred square miles on both sides of the highway. In the sky overhead I saw a white blimp, which the Department of Homeland Security was using to spot people illegally crossing the border from Mexico—one of a number of surplus reconnaissance blimps that the Pentagon had made available to government agencies once it began withdrawing American soldiers from Afghanistan. Just past the turnoff I stopped at a display of tanks, howitzers, artillery shells, and other plus-size weaponry—including an "atomic cannon," which the Army developed in the 1950s to fire tactical nuclear projectiles. I also passed a red-and-white helicopter mounted on a pole.

The road to the visitors' center runs past a cluster of RV and mobile-home campgrounds on the river and on Martinez Lake, a semi-man-made riparian lagoon. One of the campgrounds is part of the Yuma Marine Corps Air Station and is restricted to active and retired military personnel. (There are more than 250 military recreational facilities in the United States—at least one in every state but Connecticut, Iowa, Pennsylvania, and Vermont. The Air Force has a frequent-camper program.) At another campground, on my left as I turned toward the visitors' center, most of the mobile homes were permanently installed on broad concrete pads, and they looked so neat and orderly that they could have passed as self-storage units. At the marina at Fisher's Landing, just below Martinez Lake, I saw the *Colorado King*, a fifty-seven-foot-long sternwheeler that offers lunch and dinner cruises, as well as ordinary tours and private parties.

The Imperial visitors' center is pyramid-shaped, and the triangular south-facing section of the roof is covered with solar panels. Inside the building, I had the displays to myself for a while—skulls, fossils, stuffed desert animals, dioramas, informative posters, lots of interesting books—and then I was joined by two couples who were camping in RVs near Quartzsite. A volunteer showed us a live tarantula in an aquarium. Outside, beyond a second bank of solar panels, I climbed a very long ramp to a covered observation deck and looked west, toward the river. And the Colorado did look like a real river there—although not like one that could have carved the Grand Canyon, since it was barely moving, and not like one that could possibly deliver five million acre-feet a year to users between the spot where I was standing and the spot, less than fifty miles to the south, where the water runs out.

THE BIGGEST DIVERSION on the entire Colorado River system is seven or eight miles downstream from the observation deck, at Imperial Dam. The dam's construction was authorized by the Boulder Canyon Project Act of 1928—the same act that authorized the construction of Hoover Dam. Its purpose is to send water to agricultural land in California and Arizona. Most of the diverted water goes to California, mainly by way of the All-American Canal, which was also authorized by the act. The farmers who first used Colorado River water to irrigate California's Imperial Valley transported the water through a canal-and-river system that ran partly through northern Mexico—topographically, the most logical route—and Congress wanted to replace that canal, called the Alamo, with one that would be *all* in *America*, to obviate any questions regarding dual sovereignty. Imperial Dam raises the surface of the river by about twenty-five feet so that the diverted water can flow all the way to its destination by gravity. The new canal, which is eighty miles long, was completed in 1942. It's operated by the Imperial Irrigation District, which also uses it to generate a modest amount of electricity. (One of the IID's mottos is "Air Conditioning the Desert.") At the southern end of the Imperial Valley, the All-American Canal feeds some of the water into another canal, which carries it to a second agricultural area, in the Coachella Valley, 120 miles to the north. Construction of the Coachella Canal was also authorized by the Boulder Canyon Project Act.

The entire Imperial Dam facility is enormous and has many complicated-looking components. Among their functions is removing as much silt as possible from the water before routing it to its various destinations. The Colorado carries a truly staggering amount of suspended solids; before the river was dammed, its silt load, as measured at a point just north of the Mexican border, was estimated to be 160 million tons a year, roughly equal to all the material that was excavated during the construction of the Panama Canal. Quite a bit of the silt in the upper reaches of the river settles to the bottom of Lake Powell and Lake Mead, but there's still plenty in the water by the time it gets to Imperial. On the California side of the facility, the water being diverted into the All American Canal flows first into the Desilting Works, a forty-acre structure whose main elements are three concrete-lined settling basins. Silt accumulates at the bottom of the basins, then is removed by six dozen rotating scrapers. It's then pumped through buried pipes into the California Sluiceway, a three-thousand-foot-long channel that used to be known as the mainstem of the Colorado River. Some years after the dam was built, Mexican farmers complained that elevated concentrations of silt in what was left of the river were clogging their own irrigation systems. In 1964, to resolve that issue, the United States widened the river channel for a few miles downstream from the California Sluiceway and excavated a pond-size settling basin near the widened channel's midpoint. Every couple of years, that settling basin has to be dredged, and the accumulated sediments are dumped on the floodplain next to the river.

On the Arizona shore of Imperial Reservoir—just east of the dam and just north of the Gila Gravity Main Canal, which carries Colorado River water from the dam to irrigation districts in southwestern Arizona and has its own desilting facility—is Hidden Shores Village, an RV and mobile-home resort. Hidden Shores has a marina, a clubhouse, a restaurant, an equestrian area, a nine-hole golf course only slightly larger than a miniature golf course, and 625 deluxe RV and mobile-home sites. Just outside the entrance, as I drove up, two young guys in baseball caps were fishing in the concrete channel immediately downstream from the diversion control gates of the Gila canal, and two middle-aged couples were playing golf on the little course. I ate lunch at the restaurant next to the marina, then walked around the RV and mobile-home areas. Nearly every site had a carport-like awning erected over its occupants' patio furniture, propane grill, motorboat, and Jet Ski.

It was mid-December, and there were many Christmas displays, some of them quite elaborate. I saw an artificial sapling with artificial white blossoms on the branches and a big pile of fake wrapped presents stacked at its base; many Rudolphs, Santas, Frostys, and penguins; a banner, hanging from what looked like a red-roofed birdhouse mounted on a stand, of Snoopy dressed as Santa; a white stuffed dog wearing a Santa hat and standing on a red cushion in a white metal sleigh; two small octagonal signs that said "Santa Stop Here!"; a red five-gallon joint-compound bucket partly wrapped in Christmas wrap; a fake saguaro cactus wearing a Santa hat; an elaborate dashboard display featuring two Santas, several snowmen, an angel, a polar bear wearing earmuffs, and a photograph of two young children—presumably grandchildren; and a large banner on which Santa and his sleigh and reindeer were airborne above half a dozen snug-looking houses. I also saw a resident who was picking up pieces of gravel from the asphalt driveways between rows of parked RVs, and a homemade sign in front of a doublewide that said "One Old Buzzard Lives Here with One Cute Chick."

14.

IMPERIAL VALLEY

I followed the All-American Canal west to Brawley, California, one of several small cities in the Imperial Valley, and met Lawrence Cox, a second-generation farmer. He's in his late fifties, and on the day I visited his farm he had a scruffy grayish beard that didn't necessarily look permanent, and he was wearing sunglasses, a khaki work shirt, and a camouflage baseball-type cap. Both his grandfathers were airline pilots during the early years of American commercial aviation, one for the airline that became American Airlines and the other for the airline that became TWA. When his father was thinking about farming in the valley, in the early 1950s, his own father told him that he had flown over the area many times, going one direction or the other between Los Angeles and Tucson, and that it looked pretty green. The idea of farming there at all had first arisen in the late 1800s, when the ground wasn't a bit green; in fact, the entire area looked like what it was, an especially arid portion of the Sonoran Desert, and was referred to by some people as the Valley of the Dead. But the soil was flat and deep—it consists to a great extent of what's missing from the Grand Canyon—and the climate was conducive to year-round cultivation. The only missing element was water. The first farmers irrigated with groundwater, but there wasn't enough. When it ran out, they built diversions from the Colorado, fifty miles to the east. Today, the Imperial Valley is the largest single user of Colorado River water. It's also one of the most productive agricultural areas in the United States. If you eat fresh fruits and vegetables during the winter, you eat produce from the Imperial Valley.

I visited Cox in late November. There had been a little rain in Brawley on the morning of my visit, and I told him that maybe I deserved credit for bringing it—my contribution to ending the drought. "We don't plan for rain here," he said. "Our average rainfall is something like 2.85 inches a year, and everything is irrigated, and we don't dryland farm, so if we do get rain it disrupts our normal activities." (In dryland farming, limited rainfall is the only irrigation source, and the crops are ones that don't need much of it, like winter wheat.) We got into his truck and went out to see some of his fields. He and his siblings own about thirty-six hundred acres, and he leases several hundred acres more. He parked next to a concrete-lined irrigation ditch, which had slanting sides and was maybe three feet deep and six feet wide at the top. He reached into the backseat and grabbed a handful of papers from a pile the size of an unabridged dictionary.

"Here's part of our lettuce schedule," he said, running a finger down a tightly spaced spreadsheet. "We started planting September 21, and that block will harvest December 7. And we go all the way down here, all the different varieties of romaine, green, red, butter, hearts. I spend a lot of time working on this—and it's sequential." There were thirty-four blocks listed on that page, most of them ten or fifteen acres. The lettuce varieties were Javalena, Dover, Showtime, Midway, Grizzly, Bubba, Coyote, and Navajo. The planting dates stretched almost to Christmas, and the last harvest date was March 22—and there were other pages. He pulled a second sheet from the pile, with labeled rectangles drawn in pencil. "This is a market-onion map. We've got Mata Hari, Amadori, Gabriella—different varieties, different colors, different directions. Those are market onions coming up over there—as opposed to dehydrator onions, which go into onion salt and things like that. Those will be harvested in April and May."

Cox went to work on his father's farm when he was seven. His first assignment was thinning cotton, and he did such a poor job, he said, that he was demoted to weeding. He majored in crop science at California Polytechnic State University in San Luis Obispo, and after he graduated he spent five years working for his father full-time. "That was a tremendous education," he said. "My dad was on the board of the irrigation district, and he was one of the first farmers down here to have a sprinkler system, instead of flood-irrigating. He spent a lot of time on soils and salt and water efficiencies, and he was a big believer in conservation." Cox and his wife raised their children—they have two sons—in a double-wide trailer on the farm, but they moved into Brawley a few years ago, mostly because Cox got tired of being on call around the clock. "But my wife would still prefer to live out here," he said.

We drove past a plowed field that looked as level as a lake. When the Colorado was still a wild river, during the millions of years that preceded the 1930s, the course of its southernmost section shifted back and forth across southern California, southwestern Arizona, and northern Mexico, and as it did it dumped sediments over an enormous area, then planed it flat. When farmers settled there, they graded the soil flatter still, and smoothed out dunes and mounds and other irregularities, and sloped their fields to minute tolerances so that they could run water across them, slowly, by gravity alone. They also created what amounts to a valley-size plumbing system, which consists not just of canals and ditches to deliver the water but also of a network of subterranean drainage lines. The drains draw excess water from the fields, along with salt and various contaminants, and feed it into waste canals

and a pair of repurposed riverbeds, which move it beyond the cultivated area.

One of Cox's employees was kneeling in the dirt behind a big piece
of machinery, his head close to the ground. "He's planting lettuce," Cox
said, "and he's checking the seed depth. That thing behind the tractor
is a vacuum planter. You see that big fan on top, and all the hoses going
down? It picks up one seed at a time in a plate with a hole in it and
plants it at whatever spacing we set it for—in this case, three inches."
Once the plants have come up, they're thinned, leaving nine or ten
inches between them. "We figure that our vacuum planter saves us seventy dollars an acre, between the thinning and the weeding and the
cost of the seed." One of the principal threats to the plants is linnets—
small birds, members of the finch family, which walk down the rows
and surgically extract germinated seeds. "We used to use things called
bird bombs—which are basically firecrackers—to scare them away," he
continued, "but the Department of Homeland Security decided they
were possibly terrorist tools, so now we just have these little screamers."
The screamers, like the bird bombs, are fired from pistols. They make a
loud siren sound, but they don't explode. Airports use industrial versions to keep birds away from runways.

We drove to another field and parked near some baby lettuce plants.
"You can see where the next block begins, because the shoots are a different color," Cox said. The fields were irrigated by sprinklers mounted
on thirty-foot sections of three-inch aluminum pipe. The lines were
placed forty feet apart and fed by supply pipes roughly a foot in diameter. Cox's irrigation crews move the pipes from field to field in a tightly
scheduled rotation, and if they have to they can water thirty-five acres
at a time. The water is moved through the pipes by a diesel-powered
pump mounted on a trailer at a corner of the field; liquid fertilizer is
injected there, too, from a tank parked next to the pump. During a

field's final watering, specific furrows are left dry so that the harvesting tractor can drive down them without getting stuck or making a mess.

As Cox was explaining how all that worked, he suddenly interrupted himself. Sprinklers had been placed in the field next to where we were parked, and they were just coming up to pressure, and a pool of water, he noticed, was forming at the end of one of the furrows. "We must have a bad plug there," he said, pointing. The fields are separated by dirt levies, which serve as roads. He gunned the engine of his truck, and we raced around to the opposite side. "It was right there, by Juan's car," he said. He got out of the truck, jumped across the ditch, and turned off one big valve, then another. "I think that got it," he said. We drove back around to make sure, and then he explained the problem, in Spanish, to one of his foremen.

More than ninety percent of Cox's employees are Hispanic. He pointed toward a lone mountain, not far away, that was silhouetted against more distant mountains. It looked a little like a pyramid with a hunchback on one side. "You see that, there?" he said. "The base of that mountain is Mexico. We're twenty-five, thirty miles from the border."

GROWING FOOD in a desert may seem nutty, but there are many advantages. Frost, hail, and damaging rainstorms are far less common in the Imperial Valley than they are in other parts of the country, and the growing season is year-round. The jobs are year-round, too, and that means that employees can live in one place and don't have to be laid off after a few weeks or months. During the year before my visit, Brawley had received a little over half its average annual rainfall on a single stormy day, August 21, and other than that got just the odd sprinkle. Total reliance on irrigation is a drawback in one way, because the water has to come from somewhere, but the absence of rain is what makes precise planning possible: farmers in the Midwest don't know to the day when they will harvest the corn they hope to plant next month (weather permitting), and a single devastating storm or drought or flood can wipe them out for the year. (In my part of the country in 2016, an unusually warm winter followed by a hard freeze in February wiped out peach, plum, apple, crabapple, and other fruit crops—a disaster that local farmers called the Valentine's Day Massacre.) Summers in the Imperial Valley are so hot that farmers can scald plants if they irrigate at the wrong time or in the wrong way, but because winters are temperate they can get more than one harvest from one field in one year, and during the hottest months they can switch to crops that are less vulnerable to hundred-plus-degree heat. (On really hot days, rain can scald plants, too, in part because the heat drives oxygen from the falling water.)

As the western drought has worsened, though, and as the consequences of over-allocation have become apparent throughout the Colorado's watershed, Cox and many other lower-basin farmers have been forced to cut back. The Quantification Settlement Agreement of 2003 required farmers in the Imperial Valley to significantly reduce their water use, and Cox joined a group of farmers in the valley who sued the irrigation district over the terms. One of their objections was that, although they were to be penalized if they exceeded their new water allotment in any year, they were not going to be allowed to use other years' savings to offset future shortfalls. Cox and others argued that either they should have the right to store some unused water in Lake Mead or their consumption should be averaged over rolling ten-year periods. "I said, 'Wait a minute—if we have to pay back an overrun, why don't we get credit for an underrun?' But they said, 'Well, that's not the Law of the River.'" Cox was the only member of his family who supported the lawsuit, and his participation led to animated dinner-table debates. His father was dying at the time. "He had multiple myeloma and he knew his time was getting short," Cox told me. "He called me one day and said, 'Son, I want to have a talk.' He wasn't doing well, and I assumed we were going to have this big father-son discussion— 'I need you to take care of your mother'—but no. He said, 'I want to talk to you about water.'"

The IID prevailed, eventually. One result is that the valley's farmers were required to make up for an overrun of 117,000 acre-feet in 2012, mainly by fallowing land, despite having used 600,000 acre-feet less than their allotment, cumulatively, during a period of several years before and after. Cox, perforce, is making changes in many parts of his operation. Asking farmers to cut back while city dwellers are still topping off swimming pools and watering gardens seems unfair to many people, and not just to farmers, but one of the inescapable facts of western water use is that if you need to make major reductions in consumption you have no choice but to focus on agriculture. And fallowing isn't necessarily unpopular with farmers. It allows soil to rest, although if it lasts for too long it can cause damage to ditches and drainage tiles. Nevertheless, Cox said, the fallowing program had been disruptive, largely because the people who own a plot of land, and therefore receive compensation for fallowing, are often not the people who farm it. He said, "I can't blame a landlord for saying, 'Hey, Larry, I can take my ground and put it in the fallowing program for nine hundred dollars an acre, but you're only paying me three hundred dollars in rent, so I'm going to take it away for a couple of years and then we'll talk after that.'" When land goes out of production, farm employment falls, and businesses that sell seed, tractors, and other agricultural supplies come under pressure.

Cox pulled another piece of paper from his backseat pile. "See here," he said. "Forty thousand, sixty thousand, eighty thousand, a hundred thousand acre-feet—those are the efficiency gains that the IID has to achieve to transfer water to San Diego. It ramps up." I had assumed that irrigation water was measured with the industrial equivalent to household water meters, but in the Imperial Valley that's usually not the case. Cox's water consumption is monitored and controlled by one of a hundred or so IID employees known as *zanjeros*, or ditch-riders. The job is as old as western irrigation. In 1902, in an article about irrigation in *Century* magazine, a writer named Ray Stannard Baker (who later served as Woodrow Wilson's press secretary during the negotiation of the Treaty of Versailles) described a typical *zanjero* as "a bronzed man in overalls and a sombrero, who drives about in a two-wheeled cart, with a shovel and a long crook-tined fork by his side, and precious keys in his pocket." Baker added, "He is the yea and nay of the arid land, the arbiter of fate, the dispenser of good and evil, to be blessed by turns and cursed by turns, and to receive both with the utter unconcern of a small god." William Mulholland, as a young man, worked as a deputy *zanjero* for the predecessor to the Los Angeles Water Department. A modern *zanjero* typically makes two or three circuits through his assigned territory every day, manually measuring water flows at various points in the irrigation supply system.

We passed some workers who were cleaning out a ditch: a permanent, ongoing chore. Cox has lined most of his ditches with concrete, in many cases with financial help from the IID, but some of his ditches are still just dirt trenches—which not only leak water but also become choked with weeds. The weeds drop seeds into the water, and the water carries the seeds into the fields, and the ditches have to be cleared by hand. We parked in another section. "This is some of our citrus, here," he said. "It's grapefruit. It's been flood-irrigated in the past, but we're switching it all to micro-sprinkler." Micro-sprinklers are small jets that spray controlled amounts of water in a tight semicircle, which can be enlarged as the trees' roots spread.

Micro-sprinkling will help bring Cox's water consumption into line with the new requirements by virtually eliminating surface runoff, but

it will create other challenges. One has to do with the perverse efficiency effect that Bradley Udall described to me, since irrigating more
trees with less water turns a non-consumptive use (runoff) into a consumptive one (more grapefruits). That's an especially complicated issue
in the Imperial Valley, because runoff from farms like Cox's is the only
source of water, other than modest amounts of rainfall and the occasional flash flood, for the Salton Sea, an immense but shrinking and
increasingly threatened lake at the northern end of the valley. By the
time the Quantification Settlement Agreement is fully in effect, improvements in irrigation efficiency on farms like Cox's will have reduced agricultural runoff to such an extent that inflows to the Salton
Sea will be a fraction of what they have been in the past, potentially
compounding an existing ecological disaster. More efficient irrigation
also has a more localized impact, on the farms themselves. To keep salt
from accumulating in the soil and eventually rendering the land useless
for agriculture, farmers have to apply enough water to push the salt
down below the root zone and into the subterranean drainage network—
in effect, flushing it into the Salton Sea. As irrigation practices become
more efficient, though, that becomes harder to do.

IN 2013, the Pacific Institute published "Water to Supply the Land," a
report on irrigated agriculture in the Colorado basin. Michael Cohen—
Jennifer Pitt's husband—was the lead author. He told me: "The challenge with agriculture is that the new water demand is essentially all
urban—it's not agriculture that's expanding. But on the agriculture
side a staggering amount of irrigated land is in pasture crops and forage
crops, primarily for cattle. And that raises the question of whether
that's the highest and best use of water in the West."

"Highest and best use" is a concept whose main application has
been in real estate. Its basic assumption, historically, has been that the

highest and best use of any piece of land is the one that gives the land its maximum market value, and it's a tool used by people like property appraisers, tax assessors, and land-use regulators. In recent decades, the concept has been applied more broadly, to resources other than land and to values other than monetary ones. In the West, in particular, it has become a tool for thinking about how best to allocate water among potential users.

A closely related concept in such discussions is that of "embedded water," or "water footprint." The basic argument there is that when you sell or export an irrigated crop you're also selling or exporting the water that was required to grow it. This has been a controversial topic for some time; in 2014, *National Geographic* ran a widely cited article on its website called "Exporting the Colorado River to Asia, Through Hay." The contention was that when farmers in the West shipped alfalfa and other forage crops to customers in China, they were as good as shipping Colorado River water, too. This is a highly emotional issue for many people. Tom Kleinschnitz, the rafting-company owner I met in Grand Junction, told me that he gets annoyed every time he drives past a certain farm in eastern Utah, because the operator uses water diverted from the Green River, the Colorado's largest tributary, to grow hay, which he exports. "The only reason he can do that," Kleinschnitz said, "is that the people of the United States, the taxpayers, paid to create Flaming Gorge Reservoir. It was put up at great expense, and now it's being used for feeding dairy cows in China."

But even that issue is complicated. For one thing, agriculture isn't the only economic sector that has a water footprint. Patricia Mulroy told me, "Everything we transport, everything we trade across nations, has a water footprint. Do you know what the water footprint of your lithium battery is? Or the clothes on your body? There's a Tesla factory that's going in in northern Nevada, and it will use as much water as Carson City, which has a population of fifty-four thousand." Even in agriculture, "embedded water" is a nearly meaningless measure, since it can be manipulated to make a case against almost anything. If you add up all the water that people say goes into producing various food items—53 gallons for an egg, 660 gallons for a hamburger, 30 gallons for a can of beer—you end up with more than all the freshwater on earth. One reason is that people who make such calculations seldom distinguish between the consumed and non-consumed fractions: when the United States ships a ton of wheat to Africa, a quarter of a million gallons of water don't travel with it.

There are other complexities as well. Cox, like most farmers in the Imperial Valley, grows forage crops, and one reason he does is that many of them can survive summer heat that other plants can't; they therefore help to keep his fields producing income (and employment) year-round while also maintaining soil quality through crop rotation. Grasses are relatively easy to handle, and they yield multiple cuttings in short periods of time, and when water is scarce they can be allowed to go dormant—something that can't be done with broccoli or tomatoes or things that grow on trees. "That lettuce you see over there will harvest in February," he said, "and then we'll turn around and plant that section in Sudan grass, and the Sudan grass will harvest June, July, and August, and then we'll turn it back around and maybe go to onions or sugar beets or alfalfa." Forage crops, furthermore, can be irrigated with water that can't be used for other purposes. Cox now captures some of the runoff from his fields in retention ponds, then applies it again. "We're precluded by our contracts with produce wholesalers from putting reuse water on leafy greens, because it picks up bird droppings, which can contain *E. coli* and salmonella," he said. "But we can use it on alfalfa and wheat and Sudan grass." Grasses are also more salt-tolerant than other crops are. Cox drove me through a huge facility, not far from his farm, in which enormous bales of hay had been stacked into piles the size of dairy barns. "That's a hydraulic press over there," he said. "They'll take one of these big bales and apply two thousand pounds of pressure to it, and squeeze it down to maybe half its size, then load it into a shipping container. Those ones you see there, wrapped in white plastic, have already been compressed. And they can go even smaller—like those over there." He pointed at some bales that weren't that much bigger than the ones you might sit on in a horse-drawn wagon on Halloween. "They've been super-squished," he continued. "Isn't that incredible? You take all the air out, and you get a lot more weight in a container."

Exporting hay across the Pacific is economical mainly because the container ships that bring us everything we import from Asia—and we import a lot—are nowhere near full on the trip back. (For a fascinating explanation of how global shipping works, I highly recommend *Ninety Percent of Everything*, by Rose George, published in 2013.) Agricultural products are among the relatively few American goods that people in other countries are interested in buying, so selling forage crops to the Chinese is one of the relatively few tools we have for repatriating some of the dollars we pay to the Chinese for iPhones and flat-screen TVs. Cox drove us along a canal and past a farm owned by a company based in the United Arab Emirates. The farm is a source of outrage for some people, who hate the idea that foreigners, and especially Arabs, are using Colorado River water to grow alfalfa that will be consumed by non-American cattle. But the Emiratis have to do something with the billions we've paid them for oil, and because we're such good customers they're in a position to be extravagant. "Their general manager was actually one of my friends in college," Cox said. "They've spent a lot more money on infrastructure and capital costs than almost anybody down here could afford to do, and we've learned a lot from them." One of their biggest investments has been in an advanced subsurface drip-irrigation system, which is extremely efficient but also extremely expensive.

15.

THE SALTON SEA

Early efforts to irrigate the Imperial Valley with water from the Colorado River were frustrated by rapid accumulations of silt, which clogged the headgates at the mouths of diversions, washed over dikes and levees, and choked canals and ditches. In 1905, engineers working for the California Development Company, the main promoter of agriculture in the region, attempted to circumvent an especially troublesome section by making a new cut in the riverbank, just south of the Mexican border. But snowmelt in the mountains far upstream that year was heavy, and the swollen river catastrophically overwhelmed the opening, and fairly quickly the Colorado carved a new course for itself. Instead of flowing more or less due south toward the Gulf of California, it detoured west into the dry beds of two intermittent rivers. It swept away roads, farms, buildings, telegraph lines, and the tracks of the Southern Pacific Railroad, and as the water rushed forward it formed a thousand-foot-wide cataract, twenty feet tall, which chewed its way steadily back upstream, toward the mainstem of the Colorado. According to an account written in the early 1930s, "Almost at the instant that the channel was confined to any promising degree, the swift currents began burrowing and undercutting the obstruction of man's making. It is said that piles 70 feet in length were dislodged and swept away as fast as they could be driven into the water bed." Returning the river into its former channel took more than a year and a half, and, in the end, involved the dumping of three thousand railroad carloads of very large rocks. Meanwhile, essentially the entire flow of the Colorado—sometimes at a rate of more than 100,000 cubic feet per second—emptied into a broad basin known as the Salton Sink, a rift valley at the junction of the North American and Pacific tectonic plates, along the San Andreas Fault. The result was a lake with a surface area of nearly four hundred square miles. It's known today as the Salton Sea.

This wasn't the first time the Colorado River had flooded the Salton Sink. Until about four and a half million years ago, the Gulf of California extended more than 150 miles farther north than it does today, reaching almost as far as Palm Springs—and if you look down at the region from the air (as I did recently on a flight from San Diego to New York) you can see its former outline all the way up toward the northern end of the Coachella Valley. In time, though, the Colorado River deposited so much silt in the gulf that the northernmost portion was cut off, creating an isolated body of water six times the current size of the Salton Sea. That body of water eventually evaporated, emptying the Salton Sink, but the basin refilled every time the Colorado meandered back, as it did repeatedly. The vanished sea is usually referred to as Lake Cahuilla, after the native people who lived in the basin during periods when it wasn't submerged. The Cahuilla people built brush shelters, called *kish*es, whose floors they partially excavated as a defense against the desert heat. They made pottery and wove distinctive openwork baskets, and about a thousand of their descendants live in the valley today. There's a modern Lake Cahuilla, too; it occupies part of the bed of the old Lake Cahuilla, but it's just a man-made reservoir, not much bigger than a large water hazard, near a cluster of fancy golf courses a few miles south of Indio.

As captured water repeatedly evaporated from the Salton Sink, it left behind a thick bed of salt, and beginning in the late 1800s those deposits supported commercial mining operations along the shores of the Salton Sea. In the early 1940s, the Navy built a bombing range and a seaplane training area near the lake's southern end. After the war, hotel owners and real estate developers promoted the area as a winter resort and second-home destination. "The California Department of Fish and Game introduced sport-fishing species like corvina, sargo, and croaker," my *New Yorker* colleague Dana Goodyear wrote in 2015, in her article "The Dying Sea." "Tourism flourished, and with it marinas, angling operations, and yacht clubs. Speedboat regattas were held on the sea; people water-skied. An hour from Palm Springs, the sea even drew celebrities. Desi Arnaz came for the golf, and Frank Sinatra hung out at a boat-racing spot called Date Palm Beach." Postcards from that era show skiers on the water and sunbathers on the beaches. For a time, the Salton Sea was more heavily visited by tourists than Yosemite National Park, and there were so many boaters that one of the boat ramps had to be enlarged to fifteen lanes. The biggest spectator event was the Salton Sea 500, a powerboat endurance race that was televised nationally. Among the racers who set speed records on the lake was the bandleader Guy Lombardo.

The Salton Sea today covers about 350 square miles. Its surface is roughly 230 feet below sea level—just a few feet higher than Death Valley. It's an "endoheric lake," meaning that it has no outlet other than evaporation and seepage into the ground. It shrank steadily during the first two decades after the Colorado was re-diverted away from it—in the absence of inflows, the lake's surface drops by five or six feet a year—but beginning in the 1920s farmers in the Imperial and Coachella valleys began using the sink as a dump for irrigation runoff, and the water level rose. This continual topping off, which eventually amounted to more than a million acre-feet a year, made the Salton Sea's recreational boom possible, but, even though the lake appeared to have achieved a sort of equilibrium, the quality of the water steadily worsened, as evaporation caused the concentration of dissolved minerals and agricultural chemicals to rise.

Temperatures in the area reach 100 degrees virtually every day between Memorial Day and Columbus Day—occasionally, they hit 125 degrees—and the high heat, in combination with fertilizer residue from the farms, promoted algae blooms, which drew oxygen from the water. The heat also encouraged the spread of infectious diseases, which periodically combined with other factors to kill multitudes of fish and birds. Meanwhile, sporadic flash flooding and changes in irrigation volumes made the lake's contours mutable. Many waterfront properties were inundated intermittently in the seventies and eighties; others were stranded when water levels fell. The sand on the beaches was sometimes hard to see beneath the putrefying carcasses of dead fish and birds, and the odor was intense, especially when temperatures were high. It wasn't many years before real estate agents had stopped referring to the Salton Sea as a "desert Lake Tahoe." Today, the boom-time developments near the lake are either ghostly dystopias or thinly populated slums. The most depressing of the surviving settlements is Bombay Beach, a hundred-acre quadrant of mobile homes in which the occupied and abandoned properties aren't easy to tell apart. Salton City, on the opposite shore, contains a scattering of semi-expensive-looking houses, whose owners have hung on either because they're irrationally optimistic or because the collapse of the local real estate market has left them no alternative.

The Salton Sea is significantly saltier than the Pacific Ocean, and it contains growing concentrations of a long list of harmful substances.

Most of those substances are by-products of human activity, although some, including selenium, occur naturally in Colorado River water, and others were present in the sands of the Salton Sink when the current lake formed. During a four-month period beginning in December 1991, an estimated 150,000 eared grebes died while wintering in the area or making a migratory stop there. Necropsies conducted by the U.S. Fish and Wildlife Service didn't identify a cause, but did suggest a number of possibilities, including "interactive effects of contaminants, immunosuppression, a yet unidentified biotoxin found in the Salton Sea and/or a difficult to isolate manifestation of avian cholera." Among the substances found in worrisome quantities in dead and dying birds were selenium, arsenic, cadmium, chromium, zinc, and dichlorodiphenyldichloroethylene, which is one of the breakdown products of the pesticide DDT. Employees of what was then called the Salton Sea National Wildlife Refuge disposed of more than forty-five thousand dead grebes by burning them in an incinerator, which they operated around the clock. And there have been many other big die-offs of both birds and fish. Today, of all the fish species that were introduced to the lake during the boom years, only tilapia remain in significant numbers, and even they are close to the limit of what they can survive.

The Salton Sea represents a confounding ecological paradox. It was created by an act of engineering imbecility, and its continued existence and current condition are direct results of a sort of willful disregard for environmental consequences. Yet it's also the largest lake in California. It's one of the state's few surviving major wetlands, and it plays a unique and increasingly important role in sustaining the viability of a large number of bird species, more than four hundred of which have been identified on and around the lake. That role is threatened by, of all things, the state's commitment to being less reckless with irrigation water, since the more efficient the farmers become, the less water they send to the lake. The Imperial Irrigation District has been diverting Colorado River water directly into the lake to make up for some of those reductions, but its legal obligation to do so will end on January 1, 2018. After that, the inflows will drop precipitously.

The Salton Sea's receding shoreline poses a direct health threat to humans, too, because as the water level falls, more of the lakebed is exposed and the harmful substances left behind by evaporation are picked up by the wind. And the region does have wind: if you drive from the Imperial Valley to Los Angeles, you pass through one of the largest wind farms in the United States, in San Gorgonio Pass, just beyond Palm Springs, less than forty miles from the northern end of the lake. That wind farm is there for a reason. And if you drive east from the Imperial Valley on I-8, toward Yuma and Phoenix, you cross the Algodones Dunes, a vast, undulating ridge of sand that runs along the western edge of the Chocolate Mountains, where American soldiers trained for desert combat in Afghanistan. Much of that sand blew there from exposed portions of the Salton Sink, especially during periods when Lake Cahuilla was dry. The dunes are still growing and moving.

The full extent of the health threat posed by exposed lakebed is unknown, but 650,000 people live in the Salton Sea's "air shed," which extends as far as some suburbs of Los Angeles, 130 miles to the northwest. The population of the affected area is projected to double by midcentury, and there have already been unnerving increases in respiratory illnesses. (The towns closest to the lake have three times the rate of childhood asthma of the state as a whole.) In 2014, the Pacific Institute published a study of the Salton Sea called *Hazard's Toll.* The institute concluded: "The continued failure to protect and preserve the Salton Sea, worsening air quality and the loss of valuable ecological habitat—combined with diminished recreational revenue and property devaluation—could cost as much as $70 billion over the next 30 years." Where do you start? And then where do you go?

. . .

I FIRST APPROACHED the Salton Sea from the northeast, after driving
west across the Colorado Desert from Blythe on the Christopher Columbus Transcontinental Highway, also known as I-10. Twenty miles
out of Blythe, I passed the exit for Chuckawalla Valley State Prison (do
not pick up hitchhikers), and fifty miles farther along I spotted the Julian Hinds Pumping Station on the Colorado River Aqueduct at the
foot of the Eagle Mountains. From my car, as I drove, I could just
make out the main building and three huge metal conduits climbing
the steep slope behind it, a little over two miles north of the highway. I
took the exit and drove as far as the entrance gate, but there wasn't a lot
to see. The plant has nine big General Electric pumps, just like its sister
on Lake Havasu. They draw water from an open canal and lift it 441
feet to the entrance of a tunnel in the mountains, and from there the
water travels by gravity to Lake Mathews in Riverside, a hundred miles
to the west. From Lake Mathews, the water continues on to metropolitan L.A. Hinds is the last of the five pumping stations on the aqueduct.

A little farther along, I passed Chiriaco Summit, a truck-stop-size
outpost. It was named for Joe Chiriaco, who opened a gas station and
general store there in 1933. He had moved to California from Alabama
six years earlier, after traveling to Los Angeles to watch Alabama play
Stanford in the 1927 Rose Bowl—the last Rose Bowl to end in a tie.
He noticed the spot, on a desert rise that was known then as Shaver
Summit, while working as a surveyor for the Los Angeles Bureau of
Water and Power, and he guessed correctly that a business situated
there would become valuable once construction of the aqueduct began
and the principal road was paved. In 1942, the Army opened an
enormous complex—the Desert Training Center—whose purposes in-
cluded preparing American soldiers to fight Rommel's Afrika Korps.
The facility's first commander was George Patton, and his headquarters

were serendipitously close to Shaver Summit. More than a million soldiers passed through the training facility, which spread across three
states, before the Army shut it down, in 1944, and during that time the
ones who were stationed near Shaver's Summit were among Chiriaco's
best customers. When Patton died, in 1945, Chiriaco and his wife,
Ruth, created a small monument to him, and in the 1980s they do-
nated the land for what is now the General Patton Memorial Museum—
which, along with gasoline, diesel fuel, and nice clean bathrooms, is the
principal (and perhaps the only) attraction of Chiriaco Summit.

Just past the Patton museum, I took the exit for Box Canyon Road,
a two-lane highway that crosses an especially sandy, hot, and barren-
looking section of the desert and then winds through the Mecca Hills
Wilderness area—a region of dramatically folded topography created
by tectonic heavings along the San Andreas Fault. The canyon that
gives the road its name isn't deep, but it's beautiful, and in the polished
contours of its sandstone walls and in the sediments on either side of
the road you can see evidence of the flash floods that occasionally rip
through on their way to the Salton Sink. I stopped to read some informational signs placed by the Bureau of Land Management. One warned
me to be on the lookout for desert tortoises, a threatened species ("Getting closer than 10 feet to a tortoise can cause a defensive response of
emptying its bladder, which can be fatal to the tortoise"); another said
that if my vehicle broke down while I was fooling around in the desert
my best chance of being found was to stay put. Just beyond the end of
the canyon, I crossed a concrete-lined section of Coachella Canal,
which carries Colorado River water from the All-American Canal to
the Coachella Valley Water District, a distance of a little more than
120 miles. At the town of Mecca, near the southern end of the district,
I turned left, onto Highway 111, which is also known as Grapefruit
Boulevard, after one of the region's principal agricultural products. It
took me past the International Banana Museum (founded in 1972 by a

man who was crazy about bananas, even though they don't grow in the region) and down the eastern shore of the Salton Sea.

AMONG THE SALTON SEA'S early promoters was Ray Ryan, a heavy gambler who liked to cheat bookies and had many friends, and therefore many enemies, in organized crime. He also knew lots of Hollywood stars: Bing Crosby, Clark Gable, Bob Hope, Dean Martin, Frank Sinatra. In the 1950s, he and the actor William Holden went big-game hunting in Africa, and during part of that trip they stayed at an inn in Nanyuki, Kenya. They liked the inn so much that they bought it, then turned it into the Mount Kenya Safari Club, which became a popular destination for rich Americans, including a broad selection of shady characters. Ryan was one of the real estate developers who transformed Palm Springs from a desert Podunk into a winter enclave for the wealthy. He and a partner also made a speculative investment in waterfront property fifty miles to the south. They named their development North Shore Beach and described it in promotional materials as "The Glamour Capitol of Salton Sea." The centerpiece was the North Shore Beach & Yacht Club, whose main building was designed by Albert Frey, a Swiss-born architect who is sometimes described as the creator of "desert modernism."

Ryan died in 1977 in Evansville, Indiana. (He got into his car after working out at his health club, and when he turned the key in the ignition the car exploded so forcefully that lights in part of the city went out. His murder has never been solved, but no one doubts that it was ordered by the Mob.) I wasn't sure that his yacht club still existed, either. It was a popular gathering place during the sixties and seventies— various Marx Brothers and Beach Boys, among others, kept boats in its marina—but it was abandoned in 1984, and I'd seen photographs in which Frey's striking building had decomposed into a graffiti-covered

ruin. I passed the shell of a defunct tire-repair business and the shell of
a defunct grocery and liquor store, and at the turnoff for the yacht club
I stopped to take a picture of a sign that advertised building lots for
$500 down. The sign's faded condition made it both a somber eco-
nomic indicator and a fragment of archaeological narrative: so much
paint had worn away that I could make out words underneath, and I
realized that the sign had originally been the sign of the North Shore
Motel, which was abandoned when the yacht club was. Little was left of
the motel except an L-shaped concrete slab and the barely identifiable
remains of a tennis court.

But the yacht club, amazingly, was still standing. In fact, it looked
brand-new. The parking lot was empty except for my own car, but the
facade was graffiti-free, the windows were all unbroken and sparkling,
and the sand and gravel in the picnic area, off to the left, had been
raked. There was a fresh garbage bag in the garbage can. I learned later
that Riverside County had bought the yacht club and spent several million dollars on a comprehensive restoration, which was completed in
2010. The building now serves as a community center and a venue for
wedding receptions and other private parties—although it was locked
and unoccupied when I was there, during what once would have been
prime tourist season, and a two-day-old notice on the bulletin board
announced that the regular meeting of the North Shore Community
Council, scheduled for the day before, had been canceled. I walked
around the building and looked across the lake, a vast blue sparkling
sheet—and realized that what I was looking at was really the Colorado
River.

The long views were breathtaking: the Santa Rosa Mountains
seemed to rise directly from the opposite shore a dozen miles away. Closer to where I was standing, though, the scenery was semi-apocalyptic:
the restoration of the yacht club didn't extend very far on the lake side
of the building, and I saw lots of old tires and piles of broken chunks of

concrete, some of them covered with painted graffiti. I also saw (and smelled) a lot of dead fish. A dozen old wooden pilings stuck out of the water in what used to be the marina: all that's left of the docks where members once tied up their boats. (You can see the same view, in black-and-white, on the back cover of Linkin Park's 2007 album *Minutes to Midnight*.) A low, flat-roofed building a couple of hundred feet down the beach—the old snack shack—appeared ready to collapse.

I got back into my car. A mile and a half farther south, I pulled in to the Salton Sea State Recreation Area. The parking lot was huge and nearly empty, and the asphalt was as cracked and fissured as the surface of an Old Master painting, but the visitors' center was open. A friendly, knowledgeable man, who turned out to be from the suburb of Roches-ter, New York, where my wife grew up, gave me a tour. He told me that he was retired and that he and his wife spend their winters in Indio, fifteen miles north of the lake's northern end, just past Mecca and Coachella—"The summer is a little disheartening"—and that he vol-unteers at the visitors' center part-time. He showed me a video called "The Salton Sea: A Desert Saga." He flipped through a display copy of the *Salton Sea Atlas*, a big book that contained many fascinating illus-trations, and he explained that during the past fifteen hundred years the part of the Salton Sink now filled by the Salton Sea had been under-water about eighty percent of the time. I bought a hat. We stepped outside, and he pointed out the San Andreas Fault—a light-colored ridge not far to the east. I thanked him, then looked around by myself. I saw a dozen small shelters, each shading a picnic table, on a belt of clean sand next to the parking lot. No one was sitting at any of them. I walked to the edge of the water, a hundred yards beyond the shelters, and then up the shoreline. The beach looked the way beaches do at low tide. But the Salton Sea doesn't have tides: the water was low because the water *is* low, and getting lower. And much of what appeared to be sand was actually desiccated bits of dead fish, the vestiges of past calamities. The stink made breathing unpleasant. The temperature was only in the seventies, but the air felt thick.

Thirty miles farther south, at the town of Niland, I turned east, away from the lake, toward Slab City, a squatter community that arose in the 1960s on the site of an abandoned World War II–era Marine Corps camp. The Marines dismantled the original buildings when they left, but the concrete slabs that served as their foundations remain—hence the name. Slab City has no water supply, electric service, sewage treatment, or trash collection, but it has evolved into a semi-seasonal community of drifters, vagrants, unreconstructed hippies, and penny-pinching boondockers. Just outside the entrance is Salvation Mountain, a hill-size shrine that was created, over more than two decades, by Leonard Knight, a wandering easterner who lived in an old school bus near its base. He constructed it from mud, sand, adobe, tree branches, bales of straw, and tens of thousands of gallons of house paint, in many colors. He topped it with a white wooden cross and decorated it, profusely, with flowers, Bible verses, and inspirational sayings. You can climb over, around, and through it, and you can leave a donation. In 2002, Barbara Boxer, California's junior U.S. senator, described Salvation Mountain as "a unique and visionary sculpture" and "a national treasure."

In Slab City itself, I saw several dozen RVs, trailers, tents, and improvised shelters. Some presumably belonged to seasonal visitors, but most appeared to be permanent or semi-permanent installations. I passed the Slab City Christian Center, which conducts services and Bible-study sessions several days a week, and the Range, a live-music venue constructed partly from old school buses. I saw a small cluster of solar panels, a sign advertising a "Yard Sale," and the rear half of a pickup truck, which had been turned into a trailer. I also saw lots of sand, dust, abandoned junk, weathered plywood, and scrubby vegetation flecked with windblown litter. Lack of water isn't the only challenge of living in Slab City, but it's a serious one. The encampment backs up to the Coachella Canal, and many residents poach Colorado River water for bathing, cooking, and cleaning, but the nearest truly trustworthy sources of drinking water are in Niland. In Calipatria, which is one town south of Niland and is the home of the "World's Tallest Flagpole," I passed an Aqua 2000 vending kiosk, a freshwater ATM that dispensed salt-free drinking water at twenty-five cents a gallon—an unaffordable luxury for most permanent inhabitants of the Slab.

MICHAEL COHEN—the lead author of the Pacific Institute's 2014 report on the Salton Sea—told me, "There's a lot of indifference to the Salton Sea, because it's not viewed as a natural ecosystem. But for the birds that fly over that's an arbitrary distinction. They see water in the desert and they use it, regardless of where it came from." The lake is one of the principal stops on the migratory route known as the Pacific Flyway, which extends from Patagonia to the Arctic Circle, and roughly sixty percent of the species known to breed in North America have been spotted there.

Cohen earned a bachelor's degree in government from Cornell, then a master's in geography, with a concentration in resources and environmental quality, from San Diego State University. He has written three Pacific Institute reports on the Salton Sea and two on the Colorado River Delta, and he has contributed to a number of other reports on water use. He also served on the Salton Sea advisory committee of the California Natural Resources Agency. He told me, "One of the challenges with the Salton Sea—which I think is emblematic of a lot of the water problems in the West—is that any effort to protect it is really an effort to stop natural processes, and that requires a project with inputs in perpetuity. But doing that has virtually no constituency, and as a result we have a major impending ecological crisis that is ignored by almost everyone." Environmentally problematic measures that sound green (favoring locally grown food, switching to electric cars) are vastly easier to sell than environmentally useful measures that sound like trouble (dumping agricultural runoff into a reeking man-made desert lake).

Among the Salton Sea's most committed advocates was Sonny Bono, who, after the end of his recording and television careers and his divorce from Cher, served as the mayor of Palm Springs and then as a member of Congress. He died in 1998 from head injuries he suffered in a skiing accident near Lake Tahoe, and later that year the Salton Sea National Wildlife Refuge was renamed in his honor. I stopped at the refuge's visitors' center and climbed to the top of a wooden observation structure. I saw many birds and a few rabbits. I also saw Rock Hill, the eroded remnant of one of five smallish volcanoes near the southern end of the lake. (They last erupted a couple of thousand years ago.) The entire Salton Sea area is underlain by a large magma dome. CalEnergy Generation, a company that's based in Calipatria and owned by Berkshire Hathaway, operates several geothermal plants on the lake's southern shore. I drove past one of the plants and can report that it was *much louder* than I ever would have guessed a geothermal plant would be.

The Salton Sea refuge was established in 1930 as a protected breeding and wintering ground for birds and other animals. It originally included more than thirty thousand acres, but, as agricultural runoff caused the lake to grow, most of the original acreage was inundated. Today, the managed area covers about two thousand acres near the southern end of the lake, including a little more than eight hundred acres of wetlands. A large portion of the refuge consists of irrigated agricultural land sown with forage crops—including alfalfa and Sudan grass—which geese and other wintering birds feed on.

All that is threatened. "In the next fifteen years," Cohen writes in *Hazard's Toll*, "the volume of water flowing into the lake will decrease by about 40%, the Salton Sea's surface will drop by twenty feet and its volume will decrease by more than 60%. Salinity will triple." The lake has already become inhospitable to nearly all the aquatic species that once lived in it; as salt levels rise, their numbers will fall further, and so will those of the bird species that feed on them now. "The lake," he writes, "faces catastrophic change, driven most immediately by a massive water transfer between Imperial Valley and San Diego County and a subsequent reduction in flows to the Salton Sea, as well as by declining inflows from Mexico, changing agricultural practices, and a hotter and drier climate."

In 2003, as part of the Quantification Settlement Agreement, California agreed to assume all future legal liability for the Salton Sea beyond a relatively modest fraction borne by the irrigation districts. Cohen told me that health costs alone could be astronomical, since by 2030 an additional hundred square miles of lakebed will be exposed, and that there are also likely to be impacts on real estate, recreation, and agriculture in an area extending far beyond the shore of the lake. Even the most expensive of the proposed mitigation plans would be less expensive than the potential liability, but all the numbers are so large that no one who has to face voters has been eager to discuss them, much less to implement remedies. Still, Cohen told me, the liability issue may ultimately hold the key to doing something substantive about the lake. "The state is on the hook for a tremendous amount of money," he said, "and ultimately I think that's going to have more political traction than people's concern about birds."

A number of solutions have been suggested. Among the costlier ones is importing water from the (less salty) Pacific Ocean, an idea that has been around for decades and is usually the first one suggested by non-experts. "The only practical, long range solution is to dig a sea level canal from the Laguna Salada in Baja California to the Salton Sea," Fred Dungan, a decorated World War II Navy pilot and the author of self-published books on an impressive variety of topics, wrote in 2011. "This would keep the sea from getting any saltier and would also provide inland Southern California with a convenient port for international shipping." (Dungan, who also owned property on the Salton Sea, explained his proposal in his book *Bushwhacked*.) Others have suggested bringing water from San Diego Bay, eighty miles to the west, by means of a pipeline under the mountains.

Neither scheme would solve the problem, Cohen told me. "Current Salton Sea inflows are about ten percent as salty as ocean water—and about one half of one percent as salty as the water already in the Salton Sea—yet the sea's salinity continues to increase, because of evaporation," he said. "For ocean water to dilute the sea's salinity, the inflow volume would have to be very high, much higher than the evaporation rate, and that means its level would have to rise in perpetuity. It's a bit counterintuitive—it seems as though thirty-five-parts-per-thousand ocean water should dilute fifty-eight-parts-per-thousand Salton Sea water. But, because the Salton Sea has no outlet, importing ocean water would rapidly increase its salinity." "Flushing" the lake, by also pumping water in the opposite direction, is not a plausible remedy. "Depending on what elevation you wanted the surface of the lake to be," Cohen continued, "to achieve a salinity of forty parts per thousand, you would have to annually pump about 2.1 million acre-feet back over the mountains for every 2.8 million acre-feet you pumped in. That would require constant inflows and outflows roughly equivalent to a new All-American Canal in each direction."

The plan that Cohen favors (and helped to create) takes advantage of the fact that, although inflows to the lake will drop dramatically on January 1, 2018, they will still amount to more than half a million acre-feet a year. "Hundreds of thousands of acre-feet is a staggering amount of water for any kind of environmental resource in the West," he told me. "What we've been promoting is managing some of that water by capturing it in shallow dikes and spreading it out. Six or seven years ago, the USGS, in cooperation with the Bureau of Reclamation, created a hundred acres of shallow impoundments called 'pilot ponds,' and they took some water from the Salton Sea and spread it out in the impoundments, and there was tremendous bird use. It's low-key, it's not sexy, it doesn't generate a lot of recreational revenue or economic development, but as long as there's agriculture in the valley the farmers are going to need some way to leach their soils and discharge the water. If you capture that water and spread it out, you can reduce dust emissions and generate a lot of ecosystem benefits." So maybe there's hope.

RECLAMATION

The U.S. Reclamation Service—which was renamed the Bureau of Reclamation in 1923—was established by Congress in 1902, at the urging of President Theodore Roosevelt. Roosevelt is widely thought of today as one of our country's pioneering conservationists, and in many ways he was, but to him "conservation" meant much the same thing it later meant to William Mulholland and Herbert Hoover. In his first State of the Union address, in 1901, he said, "The western half of the United States would sustain a population greater than that of our whole country today if the waters that now run to waste were saved and used for irrigation." That's waste in the Mulholland sense; Roosevelt wanted the government to make western rivers useful, by regulating, capturing, and diverting their flow. "The pioneer settlers on the arid public domain chose their homes along streams from which they could themselves divert the water to reclaim their holdings," he continued. "Such opportunities are practically gone. There remain, however, vast areas of public land which can be made available for homestead settlement, but only by reservoirs and main-line canals impracticable for private enterprise."

This was exactly the kind of recklessness that John Wesley Powell warned against at that conference of irrigators in Los Angeles in 1893. There was a quasi-theological element in the notion of *re*claiming land, of taking it *back*, as though the western deserts were in a fallen state and could be returned to their rightful place in Creation through determined intervention. Among the models were the Hohokam Indians, who had built the first desert irrigation system in North America, and the Mormons, who by 1865 had built a thousand miles of irrigation canals in otherwise non-arable land. Bradley Udall told me that, for him, an especially potent symbol of the resolve of the early Mormon settlers is the Handcart Pioneer Monument, in Tabernacle Square in Salt Lake City. It's a bronze statue that depicts a family of four traveling to the Utah Territory with all their possessions in what looks like an oversize wheelbarrow. The handcarts were designed by Brigham Young himself and were used by converts who didn't own draft animals and couldn't afford to join wagon trains. Almost three thousand settlers traveled west with them, in ten organized handcart companies, between 1856 and 1860. The weight limit per person for clothing and bedding was seventeen pounds. "You look at that monument and realized that these were hardy people," Udall said. "*Walking* across the Great Plains—and it wasn't all that long ago." Two of the early companies set out too late in the year and had been equipped with inferior handcarts, which had been assembled quickly in an effort to meet unanticipated demand. More than two hundred of their members died after encountering snow and subfreezing temperatures in central Wyoming, despite the efforts of several large rescue parties from Salt Lake City. Many who survived lost fingers, toes, and feet.

Reclamation built the first dam on the Colorado: the Laguna Diversion Dam, completed in 1909, about twenty miles upstream from the Mexican border. Laguna Dam took several years to plan and design. It presented a construction challenge because it had to rest entirely on a thick accumulation of silt, with no anchor in bedrock. Engineers eventually found a workable model on the Yamuna River in India, and the Reclamation Service, supposedly inspired by a four-armed Hindu god whose image the engineers had noticed while they were there, adopted a four-armed figure as its emblem: the swastika. At the eastern end of of Laguna Dam is an unused service bridge that has twenty swastikas cast in the concrete along either side. I've got a copy of an old photograph in which the service's flag from the same era is shown flying on a pole below the Stars and Stripes. The flag has the initials *U, S, R,* and *S* in the corners and a swastika in the center. The picture looks like a dust-jacket illustration for a novel by Philip Roth.

People who go hunting for the swastika bridge sometimes fail to find it, because what looks like the dam today is actually just a narrow spillway on the California side. The dam itself is a mile long, but it's easy to miss, because it isn't tall and the river, at that point, is no longer big enough to back up behind it. What used to be the Laguna Reservoir now usually looks like an old weedy field. I parked on the shoulder of the highway on the California bank and walked across a rickety-seeming wooden bridge above the spillway to the dam, then across the dam to the bridge. Then I walked all the way back and over the road and up a hill, where I caught up to an old woman. She had long gray hair, and she was wearing a blue-and-white bandana around her head, and she was carrying binoculars and a walking stick. Her skin was lined and deeply tanned. She told me that she had lived for many years in the Laguna Dam South RV Park, directly across the road from the spot where I'd left my car, but that in 2013 all the residents had been evicted, suddenly, by the Quechan Indian tribe, whose reservation encompasses the park and the dam. The RV park had a malfunctioning septic system, which threatened the river, and the tribe had decided to clear the site. She said that the eviction occurred during the spring, after most of the seasonal residents had returned north for the summer.

Many of those people, she said, simply abandoned their property, either because they couldn't afford to retrieve it or because their mobile homes were no longer mobile. The abandoned RVs now looked like ghost RVs, and the dusty stuff lying on the ground around them looked like ghost stuff. One trailer was strung with Christmas lights.

I wondered at first whether she herself might be a Quechan, but she said that she had been born in Indiana and had moved west and stayed because she loved the desert. "I was lucky," she said, "because I live here year-round and I was able to move into town, but I'm sad not to live in the park anymore. Town is too noisy, too dirty, and too crowded. So I come out here every day." She was wearing clear plastic gloves, and she had a bulging cloth bag slung over one shoulder, so I assumed she was picking up litter. We walked together for a little while, down a dirt road that ran alongside what looked like a bigger, more robust river than the Colorado. But it wasn't a river, I knew; it was the All-American Canal, which runs parallel to the Colorado for a mile past Laguna Dam before turning west, just north of the Mexican border, toward the Imperial Valley.

MEXICO'S RIGHT to use water from the Colorado didn't become official, as far as the United States was concerned, until 1944, when the two countries signed a treaty that set Mexico's annual entitlement at 1.5 million acre-feet, to be provided either from surplus flows north of the border or, if those were insufficient, from equal contributions by the upper and lower basins. Prior to that time, many Americans believed that Mexico had no legal right at all to water from the Colorado, since none of the river's tributaries originate there (an argument that, understandably, had few supporters in California). The allocation that the countries agreed to in 1944 was just a third of what Mexico had originally wanted, but it was double what the United States originally offered. The same treaty also divided the Rio Grande: roughly two-thirds for Mexico and one-third for the United States.

Until Glen Canyon Dam was nearing completion, in the mid-1960s, large volumes of water in excess of Mexico's agreed share routinely flowed over the border, and the extra water was sufficient to sustain the river's historical delta in something like its historical condition. That changed as Lake Powell began to fill—since while that was happening surplus water remained behind the dam rather than continuing downstream—and as agricultural and municipal growth in California and Arizona accelerated, reducing the size of the surplus. An important tenet of the Law of the River is that water use doesn't "count" unless it has been formally ordered and delivered. Water takes almost a week to travel from Hoover Dam to the Mexican border if it isn't impounded along the way. If, during that time, a downstream user cancels an order—perhaps because an unexpected rainstorm has suddenly made irrigation unnecessary, or because part of a canal system requires emergency maintenance—the water continues down the river anyway and, even if it gets all the way to its original destination without being stopped by another dam, it isn't deducted from the legal allotment of the party that canceled its request.

To begin addressing this issue, in the mid-1960s the Bureau of Reclamation created Senator Wash Reservoir, an "off-stream retention reservoir," two miles upstream from Imperial Dam. Once it was completed, canceled water orders and unexpected inflows could be kept out of Mexico by pumping them uphill from Imperial Reservoir to Senator Wash—and when downstream deliveries couldn't be made with water currently in the river the stored surplus could be released. Today, the reservoir is also popular with boaters and campers, and is bordered by a large, dusty RV area, which has long views of the surrounding desert and of various distant mountain ranges. The water level looks somewhat low even when the reservoir is at maximum capacity—approximately eleven thousand acre-feet—because unrepaired structural defects in the dam prevent it from being filled to the top. When I visited, there were a few RVs parked right next to the water, on what the reservoir's designers intended to be lakebed.

In 2007, water authorities from three western states agreed to finance the construction of a second off-stream retention reservoir, plumbed directly into the All-American Canal. It's about twenty miles west of Yuma and a quarter mile north of the Mexican border, right next to I-8. It's named for Warren H. Brock, a prominent Imperial Valley farmer and agricultural innovator who died in 2006. (Among other achievements, Brock pioneered the exportation of California asparagus to Europe and Asia.) The reservoir looks like a four-hundred-acre swimming pool, and, as with Senator Wash, its main purpose is to keep canceled water orders and excess river flows from reaching Mexico. It's fed by a six-and-a-half-mile-long mini-canal, which runs mostly parallel to the All-American Canal, and it's surrounded by a tall chain-link fence that has barbed wire at the top. There's a frontage road next to the fence, and if you drive along it and then turn onto the overpass above the interstate you can see most of the reservoir in your rearview mirror. It was full when I did that—its capacity is eight thousand acrefeet, divided between two equal-size "cells," separated by a barrier. (Dividing the water makes it possible to reduce the reservoir's surface area, and hence the rate of evaporation, when it's less than full.) Intakes and outflows are managed remotely by the Imperial Irrigation District. In the scrub next to the frontage road, not far from the fence, I saw a faded sign with a Phoenix phone number, offering eleven hundred acres of desert for sale.

The Brock Reservoir cost $172 million to build. The Southern Nevada Water Authority paid $115 million, and the Central Arizona Project and the Metropolitan Water District of Southern California paid $28.5 million each. In exchange for financing the project, those three agencies will receive proportional shares of the water captured by the reservoir, above their compact allotments, until the agreement expires, in 2036. The Bureau of Reclamation estimates that Brock annually retains roughly 70,000 acre-feet—adding up to about 1.5 million acre-feet by 2036. Of that amount, Nevada is entitled to a total of 400,000 acre-feet, and Arizona and California are entitled to 100,000 each.

The Bureau of Reclamation and the participating agencies refer to this water as water that has been "saved"—and in a way that's what it is, although when people talk about conservation in this way they're really using the term as Roosevelt, Mulholland, and Hoover did, since any water "saved" by the United States is "lost" by Mexico. In 2010, a major earthquake occurred in Baja California near Laguna Salada, an intermittently dry ancient rift lake similar to (though significantly smaller than) the Salton Sink. The earthquake damaged so much water-carrying infrastructure that Mexico was unable to use its entire Colorado River allotment. Much of that water flowed over the border anyway, because there was no way to stop it, and the result was that Mexico received a little over 170,000 acre-feet "above order." That water is sometimes described as having been "wasted," but that's true only in that same old-fashioned sense, since in Mexico any part of the river's flow that wasn't used by farmers or city dwellers replenished aquifers and helped to support the increasingly threatened ecosystems of the delta. But water managers—who tend to view borders and other imaginary lines as hydrologically significant—have not traditionally thought that way.

Between Yuma and the Imperial Valley, a distance of roughly thirty miles, the All-American Canal crosses an imposingly arid stretch of sand dunes and uninhabited desert. The canal, as originally constructed, was just an enormous unlined trench, and tens of thousands of acre-feet used to seep into the ground from its channel every year. As part of the Quantification Settlement Agreement of 2003, the San

Diego County Water Authority and the California Department of Water Resources agreed to spend $300 million to line a twenty-three-mile-long section with concrete. The lining project was completed in 2010, and the Bureau of Reclamation, which owns the canal, has estimated that it reduces leakage by close to 70,000 acre-feet a year. A similar project on the Coachella Canal, agreed to at the same time, was completed in 2006; it annually prevents leakage of an additional 21,000 acre-feet. Most of the water captured by the two lining projects is allocated to San Diego, and represents a significant fraction of the water that the agricultural districts are required to transfer to municipal water districts in Southern California under the terms of quantification.

When the All-American Canal first came into use, leakage along its route caused flooding in northern Mexico, and Mexico complained. But Mexican farmers in the affected areas soon came to depend on the influx, which not only was useful for irrigation but also beneficially reduced the concentration of salt in local groundwater. In 2010, with the completion of the lining project, those benefits ended; in addition, environmentalists realized that leakage from the unlined canal had sustained a fragile Mexican wetland, which dried up when the leakage stopped. The lining project is a good example of the often paradoxical consequences of improvements in water efficiency. Eliminating leakage from the canal didn't reduce water use or turn waste into a new resource, as most people claim; instead, it transferred an existing resource from water users and ecosystems in Mexico to city dwellers in Southern California (a resource that had formerly been transferred from Arizona to Mexico), creating a shortage in Mexico that then had to be relieved with water from somewhere else.

Another unintended consequence of the lining project was that it increased the drowning danger posed by the canal—which was high already. In an influential piece on the CBS News program *60 Minutes* in 2010, John Fletemeyer, the author of *The Science of Drowning*, published in 1998, said, "I've been involved in drowning research for most of my life. And I think that the All-American Canal's probably the most dangerous body of water in the United States." The main factors, he said, are the low temperature of the water and the swiftness of the current, "which even an Olympic swimmer could not successfully swim against." The lined portion of the canal is 20 feet deep and 225 feet across, and its concrete sides slope steeply, making escape more difficult still. Most, but not all, of the people who have drowned in the canal have been people trying to cross the border illegally from Mexico. Since the broadcast was aired, several steps have been taken to make escape easier, including the addition of (widely spaced) float lines like the ones that divide lanes in large swimming pools. But the canal still poses a significant drowning hazard, and there are those who object to the implementation of any safety measures at all—in effect arguing that attempted undocumented entry into the United States ought to be a capital offense.

WATER-RELATED DEALINGS between the United States and Mexico were fairly amicable between 1944 and 1961, when they were strained by a dramatic increase in the salt content of the Colorado just before it left the United States. "By the end of 1961," April R. Summitt writes in *Contested Waters: An Environmental History of the Colorado River*, published in 2013, "salinity levels at the border had spiraled to 2,700 ppm, more than a 300 percent increase" from its concentration as measured at Hoover Dam. Mexican farmers complained—and there was little doubt about the main source of the problem, since virtually all of the increase was attributable to runoff from a single American agricultural area, the Wellton-Mohawk Irrigation and Drainage District, in the Mohawk Valley in southwestern Arizona. "While the quality had been

adequate prior to 1961," Summitt continues, "farmers now found it unusable and had to let the water flow to the delta, losing the cultivation
of about 100,000 acres."

Farmers in the Mohawk had been irrigating crops for more than a
century—first with water from the Gila River, then also with groundwater, then also with water from the Colorado River by way of the Gila
Gravity Main Canal. Groundwater in the Mohawk Valley has always
had a high salt content, and the concentration rose as farmers depleted
the underlying aquifer by drilling deeper and pumping harder. (Well-
ton's original name, in the late 1800s, was Well Town.) To keep the salt
from accumulating in the soil and ruining their fields, the farmers tried
to build a subterranean drainage system like the one in the Imperial
Valley. But because the Mohawk Valley is bowl-shaped, and because
the region contains no natural drain analogous to the Salton Sink, the
wastewater had nowhere to go, and the drainage system didn't work. In
addition, an impermeable stratum not far below the surface of the irrigated area caused excess irrigation water to accumulate just under the
fields, creating a highly saline perched system—a man-made saltwater
aquifer like the one under much of Las Vegas. And, once the perched
system existed, applying more water to the fields disastrously created a
hydrologic connection between the surface and the salty water underground, drawing salt up into the root zone and killing the crops.

To deal with that problem, the farmers began using their old irrigation wells to pump groundwater out from under their fields even as
they were using river water to irrigate those same fields. (Today, the
pumping is done by ninety hundred-foot-deep wells.) In 1961, the Bureau of Reclamation helped the farmers further, by building a concrete-
lined drainage canal to carry the extracted salty groundwater away
from their farms and dump it into the Colorado River just north of
the Mexican border. Doing that earned the Wellton-Mohawk district a

substantial return-flow credit, thereby "saving" something like 100,000 acre-feet a year, but it also caused the salt content of the Colorado River to spike just before it left the United States.

If the Bureau of Reclamation had anticipated that problem, they hadn't given it much thought. And, at first, American officials took the coldly legalistic position that the 1944 treaty dealt only with quantity, not quality, and that, furthermore, at least some of Mexico's salinity problems were self-created, caused by excessive groundwater pumping by Mexican farmers. For a number of reasons, though, that attitude changed. One reason was concern among American officials that, if the United States took a hard line on the river, an ominously left-leaning Mexican government might be tempted to ally itself with the Soviet Union. Another reason was that irrigators in the United States had salinity issues of their own, and they came to believe that, if the American government would do something to address Mexico's problem, American farmers might come out ahead, too. According to Summitt, "both western states and Mexico needed federal dollars to construct and maintain their water infrastructures, and both sides lost money to salinity damage on a regular basis."

As a result, beginning in the 1960s, the United States took a number of steps to reduce the amount of salt in the river before it crossed into Mexico, culminating in the passage of the Colorado River Basin Salinity Control Act of 1974. (The salt-removal facility in the Paradox Valley, which was built in 1991, was a direct result of the act.) The least significant salt-reduction step, so far, was also the most expensive: the construction, on the Arizona side of the border, of a large desalination facility, the Yuma Desalting Plant, which wasn't completed until 1992 and has been operated only twice, and then only to make sure it worked. (It was damaged by flooding shortly after it was completed. By that time, people had begun to question the cost of operating it while there was plenty of water upstream, and other salt-reduction measures had made it superfluous anyway—although it is still maintained and may play a role in the future.) The plant is on the western outskirts of Yuma, near the corner of Calle Agua Salada (Saltwater Street) and Reclamation Drive. I drove by one evening and wouldn't have guessed from the number of lights that were burning on the outside of the structure that it wasn't humming away. The plant also houses the Yuma area office of the Bureau of Reclamation's Water Quality Improvement Center, which tests water from the river and performs research on topics of growing regional interest, including transforming brackish water into potable water.

A salinity remedy that helped to make the Yuma Desalting Plant unnecessary was the construction, in the 1970s, of various extensions to the 1961 drainage canal. The enlarged disposal system bypasses the Colorado River entirely. It takes salty groundwater pumped out from under the Wellton-Mohawk district, carries it beyond the irrigated agricultural areas of northern Mexico, and dumps it into the desert—a change that has had profound environmental consequences that no one in either country anticipated.

THE DELTA

On a chilly Monday morning in early December, I met Jennifer Pitt in the parking lot of the Coronado Motor Hotel in Yuma, and we drove to a Mexican border crossing a few miles to the west. Part of the border in that area runs north-south, following a twenty-three-mile long reach of the Colorado River known as the Limitrophe. We walked down a covered path and through a turnstile, and emerged in the business district of Los Algodones, a town whose principal commercial offerings, as far as American visitors are concerned, are deeply discounted dental procedures and prescription drugs that can be bought without prescriptions. We had come to see Osvel Hinojosa Huerta, who is the water and wetlands program director of Pronatura Noroeste, an environmental organization. Pitt spotted Hinojosa Huerta's copper-colored pickup truck, and we crossed the street and got in.

Hinojosa Huerta is in his early forties. He's stocky, and his face is bearded but cherubic, and he was wearing a gray hooded sweatshirt with the Pronatura logo embroidered on the front. His academic training was in ornithology, and the gear in his truck included an enormous pair of truly awesome binoculars, which looked like something General Patton might have used at the Battle of the Bulge. He drove us first to Morelos Dam, just south of town. The Colorado River, as a river, essentially ends there: the dam, which was completed in 1950, diverts virtually all its water into the Alamo Canal, which feeds farms and towns in the Mexicali Valley. At the dam, the canal takes a hard right to the west, then almost immediately bends south and runs parallel to the riverbed for several miles before subdividing into lesser canals. We parked in front of a small building, took turns patting a very small dog that belonged to a dam employee, and walked onto the dam. On the downstream side was a swampy rectangle covering four or five acres. Shallow water was standing in much of it—the result of seepage through the dam, Pitt said—and the water made the rectangle look like a pond that had been drained almost all the way to the bottom. Hinojosa Huerta named a few of the birds he could see: "Belted kingfisher. Cormorant. Green heron." Then he handed the binoculars to me and pointed to a tiny break in the opposite edge of the swampy rectangle: an opening maybe the width of a city sidewalk, with dense clumps of dark green vegetation on either side. "That's the Colorado River," he said.

Construction of Morelos Dam was mandated by the 1944 treaty between the United States and Mexico. The dam is maintained and operated by Mexico under the supervision of the International Boundary and Water Commission, which was created by the two countries in 1889 to deal with a variety of border issues. Its responsibilities at Morelos include ensuring that water deliveries to Mexico comply with the 1944 agreement. It's also charged with maintaining the channel of the Limitrophe, so that an "avulsive river event"—a sudden change in the path of the streambed—doesn't accidentally shift territory from one country to the other. Fifteen or twenty years ago, Pitt said, engineers from the IBWC decided to dredge the river's channel to prevent that possibility. But their dredging plan was based on flooding studies conducted in the forties and fifties, when the river actually flowed. "I think they were just doing it because it was on a piece of paper," Pitt said. "We told them that there was an important wetland habitat down there, and we said that doing what they wanted to do would harm it, and their consultant said, 'Oh, no problem, we'll put the channel over here, away from your habitat, to protect it.' Typical engineer thinking." Pitt and others complained to the Department of the Interior, and the project was eventually dropped.

In the aftermath of the big earthquake in Baja California in 2010, Mexico negotiated Minute 318, an emergency addendum to the 1944 treaty which gave it the right to store up to 260,000 acre-feet of the unused portion of its Colorado River allotment in Lake Mead—water that it would be allowed to use later, once its damaged irrigation infrastructure had been repaired. This was an unprecedented concession by the United States, but the deal had advantages for the compact states, too, because it helped to keep the lake above levels at which various shortage-related emergency measures must, by U.S. law, be taken. Minute 318 also called for further negotiations, and in 2012 the two countries reached a more ambitious agreement, Minute 319, which affirmed the storage right, gave Mexico financial help in rebuilding and improving its irrigation systems, and established a formula by which the two countries would share the burden of future shortages.

Thanks partly to efforts by Jennifer Pitt, who was involved in the negotiations, Minute 319 also mandated the intentional onetime release of a large volume of river water into the Colorado's natural channel below Morelos Dam. That release, or "pulse flow," took place in the spring of 2014. Over the course of eight weeks, more than 100,000 acre-feet was allowed to run past the dam and through the inlet that Hinojosa Huerta had shown me. "Remarkably, we saw a beaver," Pitt said. "I was really surprised, and the beaver must have been surprised, too." For the first time in years, the Colorado River got all the way to the sea. "As time goes by, people forget," she continued. "It was cool, during the pulse flow, to see people bringing kids out to look at a river they had no idea existed." By the time the water got to the Gulf of California, it wasn't much more than a sludgy trickle, but farther upstream the experiment made a big impression. "Every day during the pulse flow it was a party," Hinojosa Huerta said.

The barriers to reaching agreement on Minute 319 were not trivial. "There were the obvious ones—language, culture, law, economics," Pitt said. "I don't know that there's any place in the world with a border that divides such an affluent country from such a poor one. I was a total novice at diplomacy before this all started, and I probably still am—but when people ask me about it I tell them there's no sort of court of water to which the United States and Mexico could appeal for a fair decision. Both countries came with the status quo, and tried to see what the other country had, in terms of assets or flexibility that could help solve their problem." Buzz Thompson, the Stanford Law School professor, told me, "It's one of the few stories in the western United States where you have a diverse set of interests coming together and reaching an agreement by which everyone ends up with something, including water for the environment."

HINOJOSA HUERTA DROVE US south on Highway 2, a shoulderless two-lane road on an embankment a dozen feet above the desert. He explained that the embankment was a levee, built many decades before to protect local residents from the river—a function almost impossible to imagine, because the channel of the Colorado was a mile to our east, and there was nothing between it and us but desert. The road was narrow, and it wasn't straight, and we passed quite a few crosses and shrines, marking the sites of fatal accidents. I looked down into the

yards of houses beside the highway, and saw chickens and dirt and old
tires and piles of brush and pieces of cars and lots of small and medium-
size brown dogs. A few miles south of the dam, Hinojosa Huerta drove
down the embankment and parked. Pitt pointed out the whitened
shells of river clams and apple snails—vestiges of a time when the river
used to flood that far—and we hiked across a few hundred yards of
desert to the channel. The river, when we reached it, looked more like a
creek, but Hinojosa Huerta explained that it wasn't even that. "At this
point," he said, "it's all just gains from groundwater"—much of which
seeps over from irrigated farmland on the Arizona side of the border.

The pulse flow didn't last very long. Hinojosa Huerta said that
much of the new plant growth that occurred immediately following it
had died, but that positive effects were still apparent, and that some
existing vegetation had been given a boost by it. Nevertheless, many of
the big plants we saw near the river were salt cedar, or tamarisk, an
aggressively invasive species that has become the dominant form of vegetation along many sections of the river, in both Mexico and the United
States. It was introduced from Asia in the nineteenth century, usually
for "erosion control." It has no native predators on this continent, and it
can live in salty, compromised soil that other plants can't, and its roots
go deeper than the roots of cottonwood, willow, and mesquite, which
are the principal large indigenous plants. It's much better than they are
at surviving droughts, and although native birds will rest in it they
won't nest or breed, Hinojosa Huerta said. If you burn it or cut it down,
it will rise again from its roots. If you remove its roots (not easy), you
have to immediately plant something else in the same spot or it will
come back. It deals with salt in the soil and groundwater in part by
drawing the salt into itself and exuding it through the stomata in its
leaves, and when the leaves fall to the ground they poison the soil at
its base, creating a defensive barrier that prevents anything else from
growing nearby. I pulled a few leaves from a bush and put them in

my mouth; they did, indeed, taste salty. Then I noticed a sort of dry
white film on the sand under one of the bushes and guessed that it was
a salty efflorescence, either from the leaves or from the soil itself. I was
about to taste it, too, but Hinojosa Huerta said that it was "probably
bird poop."

Salt cedar now covers a quarter of a million acres in the Colorado's
drainage basin. In 2001, the U.S. Department of Agriculture approved
an experiment in which relatively small numbers of another invasive
species, the tamarisk leaf beetle, were imported from Asia and released
in Colorado and Utah. Tamarisk leaf beetles eat only tamarisk leaves,
and the thinking was that they would slow, halt, or reverse the spread
of the plants. But the beetles themselves are quite hardy, since they also
lack indigenous predators, and they're now found in a dozen western
states and parts of northern Mexico. Not everyone believes that releasing them was a good idea.

Farther down Highway 2, we pulled off the road again, next to a
large diesel-powered pump on a trailer. The trailer was drawing water
from a canal. Water rights in Mexico, unlike water rights in most of the
western United States, are fungible, and aren't tied to the land. Hinojosa Huerta said that a group of NGOs had formed a trust and had
been assembling a portfolio of water rights by buying them from willing sellers for use in various restoration projects. This particular pump
was diverting water to a site a few miles farther along, on the other side
of the road in an old meander. We drove there next and saw a Pronatura pickup truck and four workers, who were using hoes and spades
to maintain some very ragged-looking irrigation furrows. Most of the
area had been inundated during the pulse flow. The purpose of the current project is to reestablish native vegetation using seedlings grown in
a nursery managed by a local farmer. The soil looked so sandy and dry
and uncooperative that to me the project seemed hopeless, but Hinojosa Huerta said that cottonwoods and willows grow very fast, as long

as they have enough water, and that once the roots of the seedlings had reached the water table, in two or three years, the plot would need to be irrigated only minimally, to maintain various understory plantings. Some areas on the same site had been restored already and were doing well. Similar efforts in equally unpromising areas, including a large site managed mainly by Pronatura and the Sonoran Institute, with help from visiting American college students, have also been successful.

Hinojosa Huerta wanted to show me the exact end point of anything that anyone could reasonably call the Colorado River—the last spot where surface water in any quantity flows along the streambed. We continued a mile or two farther down Highway 2 and pulled over again. "We'll try here," he said. He drove as far as he could across the sandy floodplain below the levee, then parked the truck. We saw coyote tracks, lizard tracks, and dried-out apple snails. He pointed out places where plants had germinated during the pulse flow, or immediately after it, but had died since then. We stopped at a test well, which scientists from both countries were using to monitor changes, post–pulse flow, in groundwater chemistry and in the level of the water table. We saw the sun-bleached skull of a muskrat.

Then we reached the river. Water was still flowing, sort of, although the stream was more like a skinny, slow-moving puddle. "We're close to the end," Hinojosa Huerta said. The actual terminus moves with the seasons, depending on the weather and the amount of irrigation. ATV tracks ran back and forth across the streambed, and there were many places where we could step from one side to the other without getting our feet wet. "So this is Mexico and that's the United States," he said, and jumped across. When I was sixteen, on a camping trip in Big Bend National Park, in Texas, I swam across the Rio Grande into Mexico, bought a pack of Mexican Raleigh cigarettes for the equivalent of about a nickel, and swam back. Doing that was easy, but doing this was easier.

. . .

ALONG THE LIMITROPHE, the tall fence that separates the two coun-tries is situated well inside what is technically the United States—in some places, more than a mile. At San Luis Río Colorado, the fence and the border take a sharp bend to the southeast, away from the riverbed, along the southern edge of the Gadsden Purchase, a thirty-thousand-square-mile piece of territory that the United States bought from Mexico in 1854. (It became part of what are now Arizona and New Mexico.) Hinojosa Huerta drove over the toll bridge, then turned left, toward the fence, and doubled back. He parked at a dead end, near a spot where an old man and an old woman were using long crooked sticks to poke at a smoldering pile of garbage. During the pulse flow, Pitt said, the water took two or three days to get this far, and by the time it arrived a crowd had gathered under and around the bridge. Peo-ple drove their cars down the bank and sat on lawn chairs next to the water. "They saw us coming around the bend in boats," she said, "and it was like this big welcome celebration. It was one of the most surreal things I've ever witnessed."

In his book *The Great Divide*, which was published in 2015, Ste-phen Grace writes, "The Colorado River Delta, once one of the planet's great wetland ecosystems, formed a maze of green lagoons that teemed with fish and was prowled by jaguars. Waterfowl filled the skies in such abundance that daylight dimmed as processions of wings passed before the sun." This was still true not much more than half a century ago. Pitt said, "The first real drying occurred around 1966. That's when the Glen Canyon Dam went up, and we started to fill Lake Powell, and as a consequence any water that wasn't being delivered to a user down-stream was accumulating behind the dam rather than flowing through Mexico to the gulf." There was almost certainly a similar drying when Lake Mead was filling, but records from that era are scarce. "There's so much space in the reservoirs now," Pitt continued, "that we're sort of in the same situation we were in the late sixties, so that, even if we have a really great water year, that water will never get here because any surplus will just be stored in the reservoirs."

We drove farther south. We saw fields in which cotton was growing, and fields in which cotton had been harvested. Cotton used to be as important a crop in Mexico as it was in Arizona, and was referred to by Mexican farmers as *oro blanco*: white gold. The town where Pitt and I had crossed the border that morning was even named for it. (The Spanish word for cotton is *algodón*.) Competition from growers in other countries has changed all that, but some farmers still plant it, as they also do in Arizona. We passed a cotton-processing facility the size of a fairground, although the enormous bales lined up outside the main structure were so dust-covered that they looked more like brown gold. We continued south, past more farms, and eventually reached the point where even the diverted water begins to run out. We saw sandy-looking fields at the outermost edge of the irrigated region, and we saw farmers burning off something—maybe winter wheat, Hinojosa Huerta said.

We stopped in the micro-village of Luis Encinas Johnson—population 199—where Hinojosa Huerta lived for three years when he was a graduate student. We were greeted by Juan Butrón Mendez, a local farmer, who then in his own truck led us across a section of desert that was so dry and salty that not even iodine plants were growing. "This is the most damaged part of the delta," Hinojosa Huerta said. He hung back a little from Butrón Mendez to avoid the dust cloud spiraling backward from his truck. Both vehicles were following a pair of bumpy, rutted tire tracks. The tracks were almost indistinguishable from the desert on either side, but Hinojosa Huerta said that if we strayed off the path in either direction we'd become hopelessly stuck. "The sand is almost fluffy here," he said, "and it's full of salt."

And then I saw our destination, a mirage-like line of green a couple of miles ahead of us: the Ciénega de Santa Clara, a 40,000-acre wet-land. The water doesn't come from the river; almost ninety percent of it comes from the Wellton-Mohawk Irrigation and Drainage District in Arizona, by way of the bypass canal that the United States built in the 1970s in response to Mexico's complaints about salt in its Colorado River allotment. Each year, the canal carries roughly 100,000 acre-feet of super-salty American groundwater over the Mexican border and dumps it. The water is too salty for agriculture, but—it turns out—not too salty to support the kinds of plant and animal life you might find in a brackish estuary.

"It's hard to believe we're in the middle of a desert," Pitt said when we'd arrived. I climbed an observation tower and looked out over a carpet of cattails and open water—an entirely different landscape from the wasteland at my back. The formation of the Ciénega was a surprise; it was discovered by Butrón Mendez, whose farm covers about seven acres. He now works part-time as a guide for people who come to camp, hunt, and fish. Half a dozen small brick cabins with white domed roofs were lined up near the spot where we had parked, and just beyond them were half a dozen thatch-roofed open shelters, plus a small water tower. Two small aqua-and-white boats were tied up at a long dock at the edge of the water. Butrón Mendez attached a car battery to a toy-size out-board motor in one of them, and we set out.

Pitt had described the Ciénega to me as "a big pancake of water." The easily navigable portions amount to only a couple hundred acres, but the entire wetland has evolved into a complex ecosystem, which attracts more than 250 species of birds. Our boat's motor was almost silent. As we glided around islands and across open water and down narrow passages between banks of cattails, we handed Hinojosa Huer-ta's huge binoculars back and forth. He named the birds we saw or heard: ospreys, American coots, marsh wrens, marsh hawks, snowy egrets, black-crowned night herons (which sound like barking dogs),

Foster's terns, least bitterns, black-necked stilts (which look as though they're walking on stilts), long-billed dowitchers, sandpipers, clapper rails (which are endangered and were the subject of his master's thesis), green-winged teals, bufflehead ducks. "Buffleheads are the nicest ducks," he said, "but they're hard to photograph because they're very shy." Sure enough, they all took off as I was bringing them into focus. Clapper rails are shy, too. The Ciénega is believed to be home to about six thousand of them, but when ornithologists do a census they have to count them mainly by listening for their calls—one of which sounds like the clapperboards that moviemakers use to label scenes and synchronize images with sound tracks. We rounded a corner and saw a flock of maybe sixty American white pelicans sitting on the water, close together, their long orange bills all pointing in the same direction. Butrón Mendez moved the boat nearer, and, as we watched, all but one of the pelicans flew off. A little later, the straggler flew off, too. Unlike the others, it flew directly over our boat, and, as it did, dropped a turd, which hit the water a few feet from our gunwale. Just beyond the spot where we had first seen the pelicans was a mini-delta of black silt speckled with white plastic trash: the terminus of the bypass canal.

In 2010, the cattails in the Ciénega caught fire, and eighty percent of the wetland burned. That was a good thing, Hinojosa Huerta said, because if the cattails aren't wiped out periodically they become too clogged and impenetrable to function effectively as habitat for most of the animals that depend on them. When the river was wild, he said, occasional raging floods would have accomplished the same thing, but fires are effective, too. A few months after the 2010 fire had burned itself out, everything was green again, and the new cattails were healthier. Like Pitt and many other people who now devote their professional lives to the Colorado River, Hinojosa Huerta came to his calling indirectly—in his case, along a highly nonlinear path beginning with clapper rails—and discovered he couldn't leave it alone.

The Ciénega is an accident, but it's the largest section of the delta that functions anything like the way the whole thing used to, and its existence shows how much can be accomplished, ecologically, with a relatively modest amount of water. Hinojosa Huerta has estimated that diverting as little as one percent of the Colorado's historical flow into the delta on a more or less permanent basis could be enough to restore 200,000 acres of wetland and reestablish a permanent link to the sea. Not many years ago, an outcome like that would have seemed impossible, but Minute 319 has raised hopes. Buzz Thompson told me, "It's only a five-year agreement, but the pulse flow was so successful that it's going to be very difficult for all the parties not to agree again." Among the beneficiaries of a truly sustained flow would be the Gulf of California. When the Colorado River stopped reaching the sea, the ecology, chemistry, and hydrodynamics of the upper gulf changed significantly. The gulf's only outlet to open ocean is far to the south, making it a "semi-enclosed sea." The damming and diversion of the Colorado raised the salinity of the upper gulf, changed the way the water circulates, and reduced sediment deposits to virtually zero—all forces that have increased pressures on indigenous species, some of which are now in danger of extinction.

An hour or so into our boat ride, Butrón Mendez turned off the motor, unhooked the car battery, and, using a pair of pliers and what looked like two unbent coat hangers, began to attach a replacement. Doing that took a while, and I had time to wonder what would happen if he couldn't get the motor going again. When we'd entered the desert, my mental list of possible outcomes hadn't included "swimming to safety." A maze of cattails stood between us and the dock, which was far away. I wished I'd brought a warmer jacket. The sun was setting, and the sky above the Sierra de Juárez, to the west, was turning from pinkish gray to grayish red. Then the motor started. By the time we reached shore, it was night.

WHAT IS TO BE DONE?

Buzz Thompson often teaches courses at Stanford about the Colorado River—including, not long ago, a seminar for adults. "They just couldn't understand how we could run a water system the way we do," he told me, "because, if you come at this with no previous understanding, everything seems puzzling." Prior appropriation, allocating a variable resource by fixed amounts, turning states into competing antagonists, basing laws on discredited science—it all seems absurd. "But then they look at the politics of it," he continued, "and they decide that we're never going to solve it."

Finding flaws in the way the western states have handled water isn't hard. Robert Glennon, in *Water Follies*, writes, "Such an allocation system creates tremendous inefficiencies. It ignores the economic value of the activity, treating higher- and lower-value uses alike. It encourages economic speculation. It creates an incentive to hoard the resource because the appropriator need not pay for the resource. The government essentially gave away the water to anyone who could use it. Most important, allocating a specific quantity of water transformed water into a commodity, like gold or timber. The prior-appropriation doctrine transformed water from a shared common resource into private property."

If you picked just about any high school civics class in the country and gave its students a year to gather information and think, they could almost certainly come up with an approach to western water use that would be more rational than the arcane patchwork we have currently. But that's not going to happen. Nor are we going to return to the mid-1800s and start again. John Wesley Powell argued that the defining unit in river management should be the drainage basin, not the state, and if his view had prevailed the western United States would have developed in a very different way. But his view didn't prevail, and whatever happens from now on will begin with the legal framework that arose instead. "As long as I've worked on western water issues," Jennifer Pitt told me, "there's always been a small but steady voice that says: overturn the compact, take down the dams, we need a revolution in western water law. But to change the law in a way that overturned or upset what people have come to regard as property rights would be very destabilizing politically." Water Buffaloes say that's a good thing because the Colorado River Compact has proven that it "works." They're wrong about that in at least one sense—no system whose central purposes include allocating nonexistent water can truly be said to work—but they're right in the sense that the compact has lasted a long time, mainly because it has usefully turned out to be less rigid than its architects meant it to be.

"We used to be very parochial, and very combative," Patricia Mulroy told me, "but that's a luxury of abundance. You can't be that way when you have shortages, in a drought. You can't take on things like the compact or priority rights or any of those other issues, because those are battles that can't be won, at least within any kind of reasonable time frame. The compact is a pyramid that got layered on and layered on and layered on, and all the pieces are interconnected—but that's also the beauty of it, because it tells me that the compact is adaptive." In recent years, the states have made some major efforts to cooperate, presumably because none of them want to truly test the complex legal structure they've erected. There's wide agreement, for example, that Nevada received a poor deal when the lower-basin states divided up their half of the river—yet Nevada hasn't pushed for a renegotiation, because to contest the existing division would be to risk ending up with less. Nevada has also voluntarily reduced its withdrawals from Lake Mead to well below its compact entitlement. In both cases, strategies that in some ways were merely protective of the status quo had effects that were arguably farsighted and broadly beneficial, at least within the context of the Law of the River.

Initiatives like those present reasons for hope. As the responsible parties ponder additional remedies, they will have many factors to consider.

OVER-ALLOCATION

The first great unavoidable truth about the Colorado River and other threatened freshwater sources is that any effective long-term program has to address over-allocation. In 2013, the Bureau of Reclamation published a chart that tracked water supply and demand in the Colorado River basin from the 1920s to the present, then projected them fifty years into the future. The bureau's median expected annual water deficit by 2060 is more than three million acre-feet. Even the best case is lousy, and the worst case—minimum supply, maximum consumption— almost doesn't bear thinking about. If everyone used all the water they have a legal right to use, there would be much less than no water left.

Over-allocation can't be addressed from the bottom—by making households and farms more efficient, or by adjusting the price of water, or by classifying alfalfa as a Schedule I drug. It can only truly be addressed from the top, beginning with a scientifically defensible determination of how much water, over the long term, is likely to be available for rational human exploitation. John Fleck—a former science reporter for the *Albuquerque Journal*, who writes an excellent blog about western water in general and the Colorado River in particular (at http://inkstain.net/fleck/) and is the author of a recent book on the same subjects, *Water Is for Fighting Over*, published in 2016—described the problem succinctly in a blog post: "There is simply not enough water in the system for everyone to take their full legal allotment. . . . People who work at the basin scale understand this. They understand that, in the long run, some sort of grand bargain (or federally imposed solution) is going to have to restrict the number of straws sucking water out of the river and the amount of water moved through each straw. But everyone working at the basin scale has to go home and face a domestic politics that is not particularly attentive to this basin-scale problem. There, people point to the pieces of paper (the Colorado River Compact, the Upper Basin Compact), and say, 'Yeah, but we're entitled to that water, *it says so right here!*'"

California, as a result of its Quantification Settlement Agreement for the Colorado River of 2003, discussed in chapter 14, is making a determined, multistep effort to reduce its annual consumption to the 4.4 million acre-feet that the Colorado River Compact entitles it to take, with a deadline of 2021. But that target, though challenging, is partly fictitious, since even in an average non-drought year there isn't enough water in the river to satisfy it if all the other compact parties take their entire allotments, too. Jim Lochhead, the CEO of Denver Water, told me, "You can only do two things. You can augment the supply of water from an outside source, or you can reduce demand, to bring it within the water budget that the river provides."

In late 2015, Utah applied to the federal government for approval to draw 86,000 acre-feet a year from Lake Powell and pipe it across the desert and up and over various mountain ranges to a reservoir near the town of St. George, roughly 140 miles to the northwest. Utah, like the other upper-basin states, has never diverted all the water that the compact entitles it to divert, and the pipeline, if built, would represent only about six percent of its legal allotment. This is the exact situation that Fleck described in his post about over-allocation: Lake Powell is half-empty and shrinking, "but we're entitled to that water, *it says so right here!*" The Utah legislature approved the project in 2006, and the state's Division of Water Resources spent eight years and $27 million preparing its federal application, which mostly concerns environmental impact. (Utah's current estimates of the pipeline's construction cost—which are almost certainly optimistic—range from just over a billion dollars to just under two billion. Las Vegas, by comparison, expects to spend $650 million for its new pumping plant alone.) The proposal will force an interesting test of the compact's viability in a time of shortage, and of the willingness of the federal government to insert itself into a matter that the states would prefer to think of as none of its business. It also poses a potentially troublesome issue for the lower basin: the water that Utah wants to divert is water that California, especially, has been accustomed to think of as its own, yet objecting to Utah's proposal would necessarily weaken California's compact claims—and, if California feels threatened, then Arizona (the junior rights holder, behind California in the line for water) has to feel even more threatened. Utah hopes to begin pumping by 2025.

Among the many difficulties of truly dealing with over-allocation, once the compact states, the Department of the Interior, and Congress have reached the point where doing so is unavoidable, is that no one can say authoritatively how much water truly exists to be divided, beyond the very short term: it's not a fixed, unvarying amount. And "normal" fluctuations will be exacerbated by climate change, whose most alarming likely effects include declining precipitation in the mountains that feed the river. Current estimates of climate-related reductions in the Colorado's annual flow, by mid-century, range from about ten percent to about thirty percent. If those estimates turn out to be accurate, existing allocations will be even less meaningful than they are now. And, if the current drought lasts even half as long as any of the extended dry periods of the past, none of the current numbers will mean anything.

Dividing the Colorado's watershed into two basins and treating them as equals provided a temporary resolution to a fraught negotiation a century ago, but in an era of permanent scarcity it doesn't make practical sense. The upper and lower basins have significantly different climates, altitudes, topographies, growing seasons, economies, populations, and patterns of settlement, and acting as though they don't makes long-term solutions harder to reach. Viewing Kent Holsinger's parents' cattle ranch in northern Colorado, the Webbs' vineyard in western Colorado, and Larry Cox's lettuce and onion fields in southern California as equivalent claimants on irrigation water is irrational if the goal is to make the best use of a threatened resource. "If John Wesley Powell had been around," April Summitt writes in *Contested Waters*, "he would have been surprised that the upper basin eventually got *any* irrigation projects." Had the seven compact states been prescient in 1922, they might have devised a mechanism for adjudicating disparities as settlement patterns and other circumstances in the West inevitably changed. Instead, each did what it could from the outset to quickly establish as many "beneficial uses" as possible—essentially what Utah now hopes to do, belatedly, with its proposed pipeline to St. George. Summitt continues, "Both the upper and lower basins were too busy grabbing what water they could instead of seeking a sustainable relationship with that water."

But the situation is far from hopeless. As Buzz Thompson says, Minute 319 shows that traditional antagonists are capable of negotiating complex agreements in which all parties acquire something they want. The usual tool for peacefully solving conflicts like this is money—in this case, probably lots of it.

THE ROLE OF THE FEDERAL GOVERNMENT

Water users in the West tend to view the federal government as a sinister, intrusive force, but the reality is that without enormous investments by all U.S. taxpayers there would be very little for the Law of the River to be the law of. Almost all the major water-storing and water-moving infrastructure in the West was financed by the federal government, whether directly or indirectly, and that means that almost anyone who uses almost any water for almost any purpose is a beneficiary of federal largesse. Water customers in Southern Nevada have paid for the intake they currently use to draw water from Lake Mead, and they're paying for the intake that will allow them to draw water from below the lowest levels accessible to Hoover Dam, but Lake Mead itself wouldn't be there if the federal government hadn't created it. Groundwater irrigation in many parts of the West was made possible by FDR's Rural Electrification Administration, which beginning in the late 1930s gave western farmers and ranchers the ability to inexpensively tap aquifers they couldn't have accessed otherwise. Farmers and other users have legitimate grievances against state and federal regulators—and Larry Cox, among others, described a number of them to me—but the water they're fighting over would not be accessible to most of them without more than a century's worth of large-scale public investment.

The antipathy of many people in the compact states to what they think of as government meddling has a public benefit, however, because it gives them an incentive to cooperate preemptively. If Lake Mead falls below 1,075 feet above sea level for a certain period of time at particular times of the year—an elevation that it first dipped under briefly in late 2015—a variety of federally mandated emergency reductions kick in. To keep that from happening for as long as possible, California and Nevada in recent years have left water in the lake which they were entitled to divert, and other states have taken similar steps. In any federal shortage declaration, Arizona would have to cut its share of Colorado River water by a significant amount, because the bargain it made with California in 1968 to ensure federal funding of the Central Arizona Project made it the lower basin's junior user. But California doesn't want Arizona to suffer so much that Congress or the Department of the Interior would be tempted to redraw that bargain. All the lower-basin users (plus Mexico) have reasons to forestall a shortage declaration for as long as they can, even if the Law of the River says they wouldn't suffer. And they all have reasons to find less punitive alternatives to agreements already in place.

DIVERSION FROM OTHER RIVER SYSTEMS

When the Central Arizona Project was being debated decades ago, it was widely assumed by serious people that, once Arizona began drawing its full share under the compact, the Colorado River system would have to be augmented with a pipeline or aqueduct linking it to a region with a water surplus—perhaps the Pacific Northwest, or even Minnesota. Such ideas have fallen from favor (Jennifer Pitt told me that people who believe in their viability as a strategy are "from Mars"), but they're still around. In 2015, I spoke by phone with Donald Trump, who at the time was only number two in the polls among the likely

Republican candidates for president. We were talking about something else (a municipal golf course in the Bronx that he currently operates for the city under a twenty-year lease), but our conversation—after I told him that I found his popularity among voters "frightening"—inevitably touched on national issues. I mentioned water in the West.

"You know," he said, "you could build a pipe, and you would never have a problem in Los Angeles and Southern California again."

"It would have to be a big pipe," I said.

"Wouldn't have to be that big. A lot of water would come down, I can tell you that. They have nothing but water."

The chance that the affected states and the federal government would ever agree to the construction of a pipeline or an aqueduct from, say, the Columbia River to the Central Valley, or from Lake Superior to southern Wyoming, seems too far-fetched to discuss, if only because the cost would be astronomical. (The states and the federal government haven't even been very good at maintaining existing water infrastructure, many elements of which are now so old that they're leaking or falling apart.) At least somewhat more plausible would be variations on the types of paper-water transfers that Nevada has negotiated with California and Arizona. Buzz Thompson told me, "Pat Mulroy once talked about an arrangement in which you would take water from, say, the Great Lakes and move it down the Missouri, and then Kansas could take some of Missouri's water, and Colorado would take some of Kansas's water, and so on. You wouldn't pipe it the whole way; you would sort of wheel it, one state at a time." The advantage of such a scheme would be that it wouldn't require much in the way of new infrastructure; the disadvantage would be that it would require the cooperation of entities with conflicting interests and water problems of their own. But, on some scale, arrangements like this may be inevitable.

"Or you could do something with the sea," Trump said, "with the—what do they call it?"

"Desalination," I said.

"Desalination. Which gets expensive, by the way."

DESALINATION

The main reason desalination gets expensive is that the equipment is complicated and operating it consumes lots of energy. The principal technology is reverse osmosis, by which saline water is placed under sufficient pressure to force it to do something that nature doesn't want it to do: pass through a semipermeable membrane in such a way that all the dissolved salt stays behind. Until fairly recently, the only desalination plant in the United States that processed ocean water was in Tampa. I was given a tour in 2011. It's an impressive facility, but it costs so much to run that it's used just part of the year, and on an annual basis it supplies only about two percent of the region's water. An official of Tampa Bay Water, the local utility, described the region's water sources to me as "a three-legged stool": groundwater, surface water, and desal. It was a one-legged stool—all groundwater, the least expensive source—until population growth and saltwater intrusion into the well fields forced diversification, beginning in the 1990s. River water, the second leg, is more expensive than groundwater, because making it potable involves more purification and processing. And desalination is the most expensive of all.

The Tampa desalination plant took a long time to build and make operational, partly because turning seawater into drinking water is difficult and partly because the local water utility and its contractors screwed up in a number of expensive ways, several of which involved large teams of lawyers. And the plant, once it was completed, had to be shut down almost immediately, when its operators discovered the hard way that it wasn't capable of handling water directly from the bay.

"Desal is a pretty simple process, except for one thing," a Tampa Bay Water official told me. "You have to make sure the water is clean before you remove the salt, or the membranes will get clogged." Fully repairing and reconfiguring the facility took until 2007 and involved the installation of an improved, multistep pretreatment system for removing silt, pollutants, algae, seashells, sewage, and other impurities. "That intake water is stripped of everything except salt and water," the official continued. The stripped water is forced through two sets of reverse-osmosis membranes, then conditioned again, by adding minerals and chemicals to "stabilize" it and make it palatable. The plant is fully functional now, although it's usually idle, as it was during my visit. When it's operating at capacity, it produces about 25 million gallons of potable water a day, or roughly ten percent of the entire service area's typical daily consumption. The extracted salt is returned to the bay, after being blended into cooling water discharged from the city's Big Bend Power Plant, next door.

The San Diego County Water Authority recently became the second American water utility with a seawater desalination plant, in Carlsbad. That facility has twice the output of Tampa Bay's and is expected to meet roughly eight percent of its service area's projected water demand by 2020. Whether that's a lot or a little depends on your point of view; it's equal to about 1.25 percent of California's Colorado River allotment. A 2010 agreement between the United States and Mexico required the International Boundary and Water Commission to "explore opportunities for binational cooperative projects that . . . generate additional volumes of water using new water sources by investing in infrastructure such as desalinization facilities." This commitment was affirmed two years later, in Minute 319, which specifically cited two potential plant sites, both in Mexico: Rosarito, in Baja California, and at the edge of the Sea of Cortez. Under one scenario, the United States would pay for the construction of a desalination plant in one of those locations, and in return Mexico would give up some portion of its Colorado River allotment.

In 2014, on a *New Yorker* assignment unrelated to water, I took a short trip on the world's second-largest cruise ship, Royal Caribbean's *Oasis of the Seas*, which can carry more than six thousand passengers and twenty-one hundred crew members and has approximately the same displacement as one of the Navy's largest aircraft carriers. Almost all the freshwater used on the ship is produced on board from seawater, using two reverse-osmosis units, each of which can produce 400,000 liters of freshwater a day, and four evaporators, each of which can produce 800,000 liters—altogether, more than a million gallons a day. The units are surprisingly small, and at the scale of a cruise ship both processes can be reasonably economical, since passengers don't travel with their lawns and dirty cars, and since they don't wash their own dishes or, for the most part, do their own laundry. Before they produce freshwater, the crew members who manage the system on the *Oasis* have to consider such variable factors as the amount of fuel on board (so that they don't reduce the ship's navigational stability by overloading or underloading the water tanks) and the time of day (so that they're prepared for sudden increases in showers and toilet flushes). The ship also processes all of its own wastewater, including an average of thirty thousand liters of sewage per day. "We mix it all together in a tank, and then we have a prescreening," the environmental officer told me. "The first concern, with toilets, is condoms." All wastewater, including graywater, is processed to the World Health Organization's standard of potability, then discharged into the ocean—a huge improvement over the once universal practice of simply dumping shipboard "black water" overboard.

Desalination is more economical if the water it processes is less salty than seawater—say, brackish water from a river estuary, or recycled water that's too salty to be reused even for irrigation, or formerly pure groundwater that has become excessively saline as rising sea levels have pushed saltwater intrusion farther inland. And there are parts of the world where desalination is the only source of freshwater—among them Dubai, which would be uninhabitable without it. The places that depend on it now tend to be places that are not just undersupplied with water but also oversupplied with fossil fuels, which the process consumes a lot of. (The electricity that runs the Tampa facility is generated by burning coal at the power plant next door.)

The number of desalination facilities all over the world is expected to rise dramatically during the next few decades, as freshwater supplies in more and more regions come under stress from population growth, over-pumping, rising sea levels, declining precipitation, and other factors. One ominous environmental consequence of that trend, with significant climate implications, is that the energy required to operate facilities that haven't been built yet is energy that the world isn't currently using. That is to say, increasing global reliance on desalination will enlarge the world's energy load, by adding a big new category to our already quite long list of energy uses—like installing central air-conditioning in a building whose occupants have always cooled themselves only by opening their windows. And doing that will inevitably increase the world's reliance on fossil fuels, either directly (if the new desal plants run on fossil fuels) or indirectly (if they run on non-carbon energy that otherwise could have been used for something else). Another worrisome environmental consequence, especially as the equipment becomes more common and therefore cheaper, is that desalination can make it possible for people to undertake environmentally disastrous development in places that couldn't have supported it otherwise—as in Dubai. Technological innovation isn't always a *solution* to environmental problems.

CLOUD-SEEDING

At the Colorado River Water Users Association meeting in 2014 in Las Vegas, Don Ostler, the executive director of the Upper Colorado River Commission, gave a presentation about the Wyoming Weather Modification Pilot Program, a cloud-seeding experiment he'd participated in. The purpose of the experiment, which ran for several years and was paid for mostly by Wyoming, was to assess the feasibility of increasing snowpack in three targeted mountain areas by shooting silver iodide from twenty-foot-tall towers on the ground into moisture-laden clouds above them. One of the three mountain areas was in the drainage basin of the Green River, a tributary of the Colorado. The hope was that the silver iodide would stimulate the formation of ice crystals, which would then fall as snow, which would eventually melt and run into the river. The conclusion of the people running the experiment was that cloud-seeding had increased precipitation in some of the test areas by between five and fifteen percent. Ostler described the technique as one of the most economical strategies for augmenting the Colorado River—although it didn't work everywhere and (to my eye, at least) the gains weren't huge.

One interesting aspect of the experiment is that Wyoming didn't stand to gain from it directly. "The storage we're trying to protect is below all our users," Ostler said. That is, the goal was to provide additional water for Lake Powell and Lake Mead, which are far downstream from Wyoming. The lower-basin states contributed money to the project, but the upper-basin states had an interest, too, because by the terms of the compact they're not allowed to deprive the lower-basin states of their compact entitlements. The full implications of that provision in a period of extended drought haven't been tested legally—the wording is ambiguous, even if the compact itself is upheld—but, once again, not even the upper-basin states are eager to subject a major component of the Law of the River to a full review in the courts, at least for the time being.

AGRICULTURE

People who live on farms that are irrigated with Colorado River water often accuse people who live in cities of causing shortages by taking water that doesn't belong to them, but throughout the western United States the main water consumers are farms, not cities. Agriculture accounts for roughly eighty percent of all Colorado River water consumption throughout the river's entire drainage basin. (Jim Lochhead, the CEO of Denver Water, told me, "We serve a quarter of the state's population and well over a quarter of the state's economic activity, yet we use only two percent of the state's water.") That means that even large cities could cut back by double-digit percentages without having much impact on total overuse, and that most serious conservation efforts necessarily focus on agriculture. But reducing water use by farmers isn't as straightforward as it may sound.

The day after Jennifer Pitt and I flew over the Colorado's headwaters with David Kunkel, a different LightHawk pilot, Andy Young, flew us to a different part of the state, well away from the mountains and the Colorado's drainage basin. Our route took us directly over downtown Denver and then south, past Colorado Springs. We were roughly a thousand feet above the ground, Young said, and from that altitude it was easy to see which people kept junk in their yards and which people didn't. I also noticed many fewer backyard swimming pools than I would have seen on a similar flight over metropolitan Phoenix or Southern California, and Pitt said that that was probably because in Colorado the swimming season is so short that the cost, including the cost of water, is prohibitive. (In much of the state, the presence of a backyard swimming pool reduces, rather than increases, the market value of a house.)

We also passed over some very dry-looking agricultural land. "This is high desert," Pitt said. "It's probably not that different from what Denver used to look like, a long time ago." The terrain beyond the residential areas was mostly brown and treeless, and it was dotted with scrubby-looking brush. The cultivated fields were easy to spot. Some were flood-irrigated with water diverted from small streams or collection ponds, and the topography of the fields was made obvious by color: dark green in low-lying sections, where irrigation water had pooled, and lighter green and brown and gray in more elevated sections, which little or no water had reached. (One of the first steps in making agricultural irrigation more efficient is to level the ground— something farmers in the Imperial Valley have spent decades doing.) Other fields, which were easily identifiable because they were perfectly circular, were irrigated with center-pivot systems, each of which relied on groundwater pumped from a well at its center. We also saw livestock-watering troughs, from which spidery lines radiated in all directions: the paths that thirsty grazing animals had worn into the ground as they converged to drink.

We saw lots more green when we reached the Arkansas River, which arises in the Colorado Rockies near Leadville and crosses southeastern Colorado, west central Kansas, northeastern Oklahoma, and central Arkansas before emptying into the Mississippi. Irrigated fields formed a broad, irregularly shaped green stripe along the river's course, and within that stripe the river itself meandered from side to side. We could also see an extensive system of irrigation canals and ditches. We flew past what's left of the Pueblo Chemical Depot, which was once one of the country's largest storage facilities for mustard gas and other chemical weapons. We flew over the town of Fowler, which lies thirty or forty miles downstream from Pueblo and has a population of about twelve hundred. In 2015, the Colorado Water Conservation Board approved a plan by which the owners of six nearby irrigated farms, who had formed an alliance called the Arkansas Valley Super Ditch, agreed to sell water to Fowler and other municipalities by fallowing some of their fields— and we could see those fallowed fields.

Approval of the Super Ditch plan came with a list of restrictions, including a requirement that no field in the program be fallowed for more than three years out of any ten, and that no more than thirty percent of any farm be dried up at one time. The purpose of the restrictions is to keep the farms functioning as farms, even as they sell off some of their irrigation water to municipal users, and to avoid one of the unanticipated consequences of some other buy-and-dry schemes: the collapse, or near collapse, of the surrounding community, as farms that once supported the local economy shut down. The fallowing of even a few farms in a single irrigation district can be lethal to the entire district if the level of irrigation falls below whatever threshold is needed to keep the ditch system functioning, and if agricultural production falls below whatever threshold is needed to maintain employment levels and farm-dependent local businesses. The approval also included a requirement that the irrigators as a group maintain the "return flows" stipulated by their water-court decrees to protect users downstream (and to maintain compliance with the terms of Colorado's compact with Kansas regarding diversions from the Arkansas). Pitt told me that Colorado is unique among the states in attempting to devise farm-to-city water-transfer programs that aren't lethal to farming.

How far to go in protecting farmers, and which ones, is among the more complicated western water issues. "Agriculture" isn't a monolith; some crops are clearly more valuable than others, even if no one agrees which crops those are. (If I were king, I'd ban growing corn for ethanol— not an issue in the Colorado River basin—long before I thought about going after alfalfa.) But even low-value crops can have high-value human dependents, as in the Grand Valley, and decisions about who survives and who doesn't shouldn't necessarily be left to water managers in big cities. Or to energy producers. Kent Holsinger told me that his parents' cattle ranch (described in chapter 2) was once nearly shut down by a series of local droughts, but that his parents had been able to remain in operation largely because at a critical moment an oil and gas company offered to lease their mineral rights for several years. "The company ended up not drilling, because that area wasn't terribly attractive, but the money literally tided us over," Holsinger told me. "Now my parents have a little next egg, and they paid off debt, and the ranch is still in the family." The larger question is whether the preservation of an irrigated cattle ranch is a good thing or bad thing. As always, the answer depends on your point of view.

WATER FOOTPRINT

People who worry about irrigation often argue that it shouldn't be used to support "low-value" crops. But economics alone is an unreliable tool for determining what should and shouldn't be grown—and, as Pitt pointed out to me, "high-value" crops include things like the kinds of thirsty decorative plants that people buy to put in their irrigated yards. How about Halloween pumpkins? Christmas trees? Wine grapes? Pears for hard cider? Organic baby carrots? People who fume about agricultural subsidies—and I have sometimes been one of them—don't necessarily recall that the beneficiaries of those subsidies include pretty much everyone, since we would all pay more for what we eat if economic protections for farmers didn't exist. One of the unintended consequences of cutting back on irrigation is that it opens up former agricultural land for sprawling residential development and other "high-value" uses, whose environmental impacts can be more problematic than those of irrigated agriculture. A new subdivision in some especially dry part of the West—say, on the outskirts of metropolitan Phoenix—will almost always use less water than any irrigated farm it replaces. But water use isn't, or shouldn't be, the only consideration.

Markets aren't even all that good at managing markets. (Think of Enron and the Great Recession.) Traditional economists tend to undervalue goods that are hard to pin a price on, like air quality, water quality, and the future of civilization—those pesky "externalities." Arizona used to be one of the world's leading producers of cotton; today, the state's cotton acreage is just fifteen percent of what it was at its peak, in the early 1950s, before state officials began to worry about groundwater. That decline is generally viewed as a good thing, both for Arizona water users and for the environment—but is it really better for humanity if almost all cotton is grown instead in places like China, India, Pakistan, and Brazil (globally, the number one, two, four, and five producers), where freshwater supplies are more dangerously threatened than they are in the United States (number three)? Or maybe humans shouldn't be growing cotton at all and should instead wear only Under Armour T-shirts, Nike yoga pants, and other clothing made from synthetic fibers. Yet synthetic fibers are made from fossil fuels—and that must be a bad thing, unless it's better to turn petroleum into clothing than to set it on fire. (These aren't trick questions; if the answers are obvious, I don't know what they are.)

The resource footprint of the increasingly finicky preferences of affluent consumers all over the world is a seldom-discussed environmental issue. Larry Cox, the farmer I visited in the Imperial Valley, told me, "The durum wheat we grow for pasta has to have a certain color and a certain protein content or there's a deduction. The lettuce we sell to a food-service company has to weigh between forty-eight and fifty-two pounds, but the lettuce we grow for a retailer has to be a certain size and can only weigh twenty-eight pounds, and the broccoli we ship to the Pacific Rim might be on a ship for eighteen days, so we have to pick it very young and put it in a controlled-atmosphere bag." Fast-food chains and other commercial buyers reject potatoes that don't meet exacting specifications for color and size; ketchup manufacturers are equally picky about tomatoes. Each of those preferences has an environmental impact.

Cox continued, "Even on the citrus we grow—lemons for export have to be different sizes from lemons for domestic. Then we've got fancy grade, choice grade, standard grade. We put up thirty-two different packs, between lettuce and broccoli and mixed lettuce and sleeved romaine—and it's okay, but you kind of feel like a poodle jumping through hoops." The gourmet infatuation with tiny vegetables has water and energy implications. So does the preference for organic produce, which, because the yields are lower, requires both more water and more land, thereby encouraging "agricultural sprawl," which the writer James McWilliams defines as "an insidious form of development that threatens the world's remaining natural resources." Commercial-scale organic farming also depends heavily, though indirectly, on irrigated forage crops, because doing without synthetic fertilizer means relying on manure, most of which is necessarily produced by livestock—and that means that the water embedded in forage crops is embedded in organic vegetables as well.

THREATS BEYOND THE BASIN

According to Patricia Mulroy, the biggest challenge to the millions of people who depend on water from the Colorado River lies far outside the river's drainage basin: in Northern California. "You are never going to find long-term stability in the Colorado system," she told me, "until you stabilize the Bay Delta. It blows people's minds when I say that, but it's true." She was referring to the Sacramento–San Joaquin River Delta in Northern California. Bradley Udall agrees with her; he described the delta to me as "the biggest potential water disaster in the United States."

The Sacramento River is the largest river in California. It arises in mountains in the far northern part of the state, flows down the center of the Central Valley and through downtown Sacramento, and empties into the Pacific Ocean by way of three connected bays: Suisun, San Pablo, and San Francisco. The San Joaquin River arises several hundred miles to the south, in mountains in the eastern part of central California, and also empties into the Pacific Ocean by way of the same three bays—but from the other side. The region where the two rivers meet is almost always referred to as a "delta"—it's known variously as the Sacramento–San Joaquin River Delta, the Bay Delta, and the California Delta—but it's actually an "inverted delta," because the topography of the Central Valley and Suisun Bay is such that most of both rivers' sediment load is deposited inland, forming an alluvial fan that seems to spread out in the wrong direction, away from the ocean. That alluvial fan is filled with closely spaced, irregularly shaped islands, and from the air the whole thing resembles an enormous green jigsaw puzzle.

Settlers began growing crops on the delta's islands in the late 1800s. As they did, the islands' soil—which consisted largely of marshy accumulations of decaying plant material—began to subside. To keep river water from inundating their fields, the farmers built dikes around the perimeters of the islands, and as their fields sank below the water table they also installed drainage ditches and pumps. Most of the islands today are deeply saucer-shaped. The cultivated fields, which altogether cover more than a thousand square miles, have subsided as much as twenty-five feet below the surface of the river channels that surround them, and the dikes and pumping systems have grown as the challenge of keeping out the bay has increased.

The Sacramento River provides a huge volume of water to Southern California. In order for that water to get to the south, though, it first has to flow into the delta. Then, at the delta's southern edge, the water is pumped into a network of canals, the largest of which, the California Aqueduct, is longer than the Central Arizona Project. The main danger that both Mulroy and Udall anticipate is saltwater intrusion. If a large influx of water from the Pacific Ocean ever pushed all the way back into the delta, which serves as an enormous north-south junction box, the most important element of the state's surface plumbing system would become unusable. "Instantly, your freshwater turns to seawater," Mulroy said. "And do you know what that does to water supply in the state of California?" One thing it does, besides creating an emergency for millions of people, is to seriously complicate California's intention to reduce its reliance on the Colorado River, as well as its recent determination to stop the over-pumping of groundwater.

Catastrophic saltwater intrusion is more than a remote possibility. "One good earthquake would do it," Mulroy said. "The delta is like Little Holland—the Little Netherlands. The farms and communities are all below sea level, and they're protected by these old earthen dikes." It's the steady inflow of river water that holds the saltwater back, preventing it from entering the freshwater distribution system: breaching the dikes would be like pulling a plug, and doing that would draw a flood of seawater toward Suisun Bay. Drought and overuse pose a similar threat by reducing the force exerted by the freshwater side of the ocean interface, and rising sea levels do the same thing from the other direction. Mulroy's preferred solution—and Governor Jerry Brown's—is to build two tunnels under the delta, each forty feet in diameter and thirty-five miles long, so that, no matter what happens overhead, water from the Sacramento can be transported to the other side, and therefore to the south, without contamination. There's been a proposal to do that for many years, but no one has started digging yet, partly because the estimated cost, including environmental protections and efforts at remediation, is close to $20 billion. And, quite obviously, the likely dollar cost isn't the only reason people are concerned about that project, since the potential environmental impacts, including damage to the habitats of endangered and threatened species, are huge. Yet losing the state's largest single source of fresh surface water would trigger a long list of other disasters.

ENVIRONMENTALISTS VS. ENVIRONMENTALISTS

Some of the fiercest arguments about climate change have been not between climatologists and "deniers" but between environmentalists and environmentalists. The problem is seldom that one group has all the facts and everyone else is deluded; as frequently happens with complex issues, the biggest impediments to effective action have been truths, not falsehoods, and the fiercest arguments have often been between ostensible allies. Hydroelectric power is substantially emission-free, but dams destroy ecosystems and human communities. Uranium-235 has a small carbon footprint, but what about accidents, earthquakes, terrorists, and nuclear waste? American gasoline is one of the cheapest manufactured liquids in the world, but taxing it more heavily would increase unemployment and might push the country back into recession. Photovoltaic panels and solar thermal concentrators have potential as electricity sources, but building utility-scale installations would devastate the deserts that are the ideal places to put them.

In the 1970s, Flagstaff, Arizona, began recycling wastewater for a variety of purposes (although not for drinking). One of those purposes is snowmaking at area ski resorts, which are a major component of the regional economy. But a coalition of Indian tribes has been trying to

stop that use for more than a decade on the grounds that spraying processed human waste onto ski slopes in the region desecrates mountain
land sacred to the tribes. More recently, the tribes have been joined in
those lawsuits by several environmental groups, which have argued that
the recycled wastewater harms ecosystems. The litigation shows not
only how tricky water issues can be—is recycling toilet flushes a good
thing or a bad thing?—but also, more broadly, how hard it is for well-
meaning people to agree on environmental best practices. For more
than a quarter century, the city of Tallahassee, Florida, recycled much
of its treated wastewater by using it to spray-irrigate a twenty-one-
hundred-acre farm southeast of the city—seemingly, an ingenious in-
stance of turning an environmental liability into an economic asset.
But a 2006 tracing study by the USGS—which used a non-infectious
but easily identifiable virus as one of the markers—proved that the
practice was responsible for sharply elevated nitrogen levels in groundwater several miles to the south, and Tallahassee had to redesign its
entire sewage-treatment procedure.

Artificial snowmaking, no matter where the water comes from, impacts the environment in other ways as well. It does so directly, because
turning water into snow and spraying it onto ski slopes requires energy, inevitably supplied primarily by fossil fuels, and also indirectly,
because lengthening the ski season, which is snowmaking's main purpose, increases the energy consumption and carbon output of a long list
of ancillary activities, including air travel, car travel, and second-home
construction. Snowmaking equipment is far more energy-efficient today
than it was when it was first used commercially, in the 1970s, but improved efficiency has made using it more affordable, thereby transforming it, within the skiing industry, from an interesting extravagance to
an economic near-necessity, with the consequence that the overall energy and carbon footprints of alpine skiing have grown. Where you

stand on these issues, assuming you aren't an aggrieved member of the Hopi Nation, is probably not entirely unrelated to how you like to spend your winter vacations.

MANAGING GROWTH

When you mention western water problems to people who don't live in the West, they will sometimes say that it's crazy for people to live in water-challenged places like Phoenix and Denver and Los Angeles. Assume for a moment that they're right; where should those people live instead? In small towns in Vermont? In condominiums in Miami Beach? If the two million current residents of Las Vegas didn't live in Las Vegas, they wouldn't vanish from the earth; they'd live somewhere else. And although relocating them en masse—assuming that were somehow possible—might make life easier for the Colorado River users who remained, it would obviously have other consequences. For them, of course, but also for everyone else.

The U.S. Census Bureau has projected that the population of the United States will likely reach 400 million by roughly mid-century—an increase of more than 100 million during the fifty years following 2000, equal to the combined current populations of the country's dozen or so largest metropolitan areas—and all those new Americans won't necessarily establish themselves in places that have abundant supplies of freshwater. In fact, the American cities with the largest absolute increases in population in recent years have been concentrated in some of the most water-threatened parts of the country, and the top ten cities include three that depend heavily on the Colorado: Los Angeles, Phoenix, and San Diego. This seems like folly, water-wise—but from a broader environmental point of view desert regions are by no means terrible places for people to live. Cooling rooms from 110 degrees to 75 degrees requires less energy than heating them from minus 25 to 68; solar exposure in deserts is often high year-round, making all forms of solar-energy harvesting more attractive; the daily solar peak in deserts roughly coincides with the daily peak in human energy demand in those same areas, something that often isn't the case in other environments; and deserts impose absolute limits on some kinds of development-related environmental damage, assuming that residents can be prevented from growing too much grass on sand. With good land-use management, water scarcity can even be a useful tool for containing the heedless sprawl of human habitation.

Unfortunately, we don't have a history of taking advantage of that fact. Jim Lochhead of Denver Water told me, "I think we have a problem, in the West in particular, but probably throughout the United States, that there's a disconnect between the water utility and the land-use departments. The dynamic is that land-use authorities approve development, and then the water utility is expected to find the water necessary to serve it, and there's no discussion about whether in fact the water is there. State laws require that you certify the existence of a water supply, but it's not a very high bar to overcome, and there's not a real conversation about land-use patterns and efficiency, and whether sprawl is really a sustainable model fifty years into the future—especially given climate change—from a water-supply perspective. We serve most of the Denver Metro area, including over forty different municipalities or districts, and we don't control land use within any of those, and there's really not a forum to discuss what's sustainable for the area as a whole."

Water managers are often at cross-purposes with people who worry about the environmental impact of (for example) suburban sprawl, because increasing the water efficiency of existing households creates a water surplus that can be used to support the construction of more sprawling subdivisions. It would make sense for metropolitan water managers and land-use planners to work together, with the goal of guiding or at least planning for growth, but, as Lochhead says, that sort of interdisciplinary cooperation doesn't happen very often—and in many cases it isn't even possible, because existing land-use ordinances and other laws don't permit it. Throughout the United States, government-level decision-making related to managing the country's inevitable growth is highly fragmented, to the extent that it exists at all, and one result is that solutions devised by one department often end up creating or aggravating problems that then have to be addressed (or ignored) by others. Even within categories, decision-making is often diffuse—so that, for example, the people who determine how much water is needed for a particular service area and what price consumers will be charged for it are not the same people who are contractually or legally obligated to supply that water.

As I drove between Boulder and Denver International Airport, I passed Candelas, a fifteen-hundred-acre residential and commercial development, still under construction, which advertises itself as "the next great place on the Front Range" and, according to a sign I saw, one that offers "life wide open." The community's water is supplied by the city of Arvada, which gets most of its water from Denver Water, which gets half its water from the Colorado River system. When Candelas is finished, it probably won't have as many acres of lush green grass and vibrant wildflowers as the pictures on its website suggest it will, but it's still a good example of the kind of suburban expansion that is made possible, if not inevitable, by both existing land-use regulations and successful municipal water conservation programs. Making people more efficient at using water isn't a gain for the environment if the gains are reinvested in sprawl.

TRUE SPRAWL

When most people, including most environmentalists, think about unsustainable population growth, they picture pop-up suburbs like Candelas, not isolated mountain communities like Redstone, Colorado.
People who live in relative solitude up in the hills, in what feels to them
like environmentally sensitive harmony with nature, don't view themselves as part of the problem. But they are. People who generate their
own electricity from rooftop solar panels haven't removed themselves
from "the grid." They remain as dependent on "the grid" as anyone
else, since the vast infrastructural network that makes their lives possible (roads, stores, highway snowplows, UPS trucks, freight hubs, airports, Amazon Prime) wouldn't be there if "the grid" didn't exist.

In 2006, Melissa Holbrook Pierson, a writer who lives in a smallish
town in the Hudson River Valley in upstate New York, published *The
Place You Love Is Gone*, a deeply felt paean to the lost American landscape, the one obliterated by sprawl. In the book she accuses New York
City of having destroyed a pastoral paradise in order to create the extensive upstate reservoir system that supplies its drinking water—of "rub-
bing its chin in contemplation of turning faraway valleys into pipes to
service its water closets." The city's early-twentieth-century planners, anticipating the population growth to come, condemned farms and rural
hamlets far from the city, just as Denver's planners did at roughly the
same time, in order to build the extraordinary chain of reservoirs without
which New York City could not exist, and Pierson describes this massive
engineering project as "larceny." Her arguments persuaded Anthony
Swofford, who reviewed the book in *The New York Times*. He wrote,
"The story of New York City's water grab is astonishing, nearly unbelievable in its scope and greed," and he described the creation of the city's

water system, as recounted by Pierson, as "rural slaughter for the survival of the city."

But this is wrong. If New York City could somehow be dismantled and its residents dispersed across the state at the density of Pierson's current hometown, what remains today of pastoral New York State would vanish under a tide of asphalt. Similarly, if you demolished all of Colorado's dams, aqueducts, and water tunnels, and spread the 4.5 million current residents of the East Slope across the state thinly enough to accommodate everyone's water needs with local streams and groundwater, you'd have an environmentally disastrous mess: the suburbs of Houston or Atlanta smeared across the Rocky Mountains. It's big, thirsty, densely populated cities that make sparsely inhabited wild places possible—a package deal. And it's the existence of places like Denver that makes the continued existence of places like Redstone possible. Pierson does write, near the end of her book, that "it is the thousands of acres of uninhabited, forested land in the buffer zones of the New York City watershed that have preserved wilderness in the midst of an inexorably creeping urbanization." But she doesn't acknowledge the role of her own form of nostalgia in the destruction of the thing she loves.

THE GREAT OUTDOORS

A long-standing but unexamined tenet of conventional thinking about the environment has been that humans must personally experience unspoiled places in order to value them, but that idea doesn't actually align very well with our history as ecosystem destroyers. Wild landscapes are less often ruined by people who despise wild landscapes than by people who love them, or think they do—by people who move to be near them, and then, when others follow, move again. Thoreau's cabin, a mile from his nearest neighbor, set the American pattern for creeping residential development, since anyone seeking to replicate his experience needed to move at least a mile farther along.

The National Park system is one of our country's most remarkable public achievements; the subtitle of Ken Burns and Dayton Duncan's six-part documentary about the parks, which aired on PBS in 2009, is "America's Best Idea." The system was established more than a century ago, and it now encompasses close to 90 million acres and receives almost as many visitors each year as there are residents of the United States. It is undoubtedly the federal program with the highest level of support among people all over the political spectrum. It's managed intelligently (although it's currently facing huge funding difficulties), and it's a paragon of democracy in that it accommodates everyone from wilderness explorers to people in wheelchairs.

As with all good things, though, the benefits come at a cost: like most of what we love best about modern life, it's heavily dependent on fossil fuels. Indeed, the rise of the national parks closely parallels the rise of the automobile, which made them accessible. Today, people who visit national parks often visit them not just in enormous RVs but in enormous RVs pulling trailers loaded with other gasoline-powered recreational equipment. Even park visitors who live out of backpacks have to get from home to their starting point, and once they've reached it they depend on many of the same energy-dependent support services that RV visitors do. The hard truth is that, in an affluent, mobile society like ours, it's easier to find fault than to be blameless.

WATER IN A LARGER CONTEXT

In 1967, voters in Boulder approved a small increase in the city's sales-tax rate and used the proceeds to buy up farmland and other open space around the perimeter of the city, and in 1989 they approved an additional tax increase for the same purpose. The city's Greenbelt now comprises more than fifty thousand acres and forms a moat-like divider between Boulder and subdivisions like Candelas, a few miles to the south—a divider you can easily see from the air, as Pitt and I did on both of our airplane trips. Over roughly the same period, Boulder has adopted a succession of measures that sharply limit new construction, and therefore population growth, inside the belt. All these initiatives are often described—and not just by people who live in Boulder—as examples of environmental enlightenment and, taken together, as a strategy for defeating sprawl. But they don't work that way at all. The Greenbelt has probably enhanced the quality of life of the hundred thousand or so people who live inside it, but it has done that primarily by preserving the city's low-density, automobile-dependent develop-ment pattern, as established by the early 1970s. (Boulder, within the Greenbelt, is more sprawling than the average American community and is more than twice as spread out as Los Angeles.) And beyond the circle the Greenbelt's only effect on sprawl has been to push it farther to the south, toward Denver, and farther into the Eastern Plains.

Not that anyone has figured out an easy way to halt sprawl. In 2009, I asked Patricia Mulroy whether metropolitan Las Vegas might not be able to ameliorate some of its water difficulties by taking steps to cap its population growth and halt its horizontal advance across the valley. She said, "Please share with me how you're going to do that. Under our constitution, everyone who owns private land has a right to develop it to its highest and best use, so controls on that end have to come from the land-use side. What I have said to this community is, 'Yes, you can continue growing, but you cannot do it the way you have in the past. You have limited water resources and you live in a fragile environment, so you're going to have to plan development that is much friendlier with outdoor water use.'"

The Great Recession put a halt to the growth of Las Vegas, but even without it there was a limit to how far the city could spread. Eighty percent of Nevada is federal land, a broad ring of which fully surrounds the metropolitan area. "Development can only occur inside that ring," Mulroy told me in 2014, "and the rest can't be developed." That's different from Boulder's Greenbelt: Las Vegas's version really does impose a sprawl constraint, because there's nothing on the other side but land controlled by the United States. "That doesn't necessarily mean you won't have growth," Mulroy continued, "but the way people live will change. When I got here, we never had high-rises like the ones you now see all over Southern Nevada. Everybody had their quarter-acre with their two-thousand-square-foot house and grass all the way around it. Well, the lots are getting smaller, the developers are getting smart, they're building a lot of common area and assimilating into the natural environs, and they're building high-rises. We are Manhattanizing, instead of becoming L.A.-like. You can already see it in the next generation."

One challenge for urban planners is that, the better a municipality gets to be about managing water use, the lazier it can afford to be about managing growth. This is a stealth issue in metropolitan Los Angeles, for example, whose residents use less water, in aggregate, than they did when there were a million fewer of them. And, although climate change has a bigger impact on water use than it does on growth, growth has a bigger impact on climate change than water use does. The world's gathering environmental problems are deeply interrelated, and they can't be addressed effectively if they're addressed in isolation: water here, energy there, transportation somewhere else. But this sort of big-picture environmentalism is hard to define, much less to pull off, since even people who worry about the future of civilization tend to specialize.

BACK TO HERBERT HOOVER

In a speech in 1926, when he was still the secretary of commerce, Herbert Hoover said, "True conservation of water is not the prevention of its use. Every drop of water that runs to the sea without yielding its full commercial returns to the nation is an economic waste." This sounds prehistoric to a modern reader, but if you enlarge your conception of value beyond the merely economic—by including the full list of "externalities," including the environmental value of allowing water to run into the sea—it's not a terrible template for thinking about human resource exploitation of all kinds. The ideal resolution to many western water issues would intelligently address other issues, too, including energy and climate, and would attempt to arrive at some degree of mutual accommodation among a long list of competing and often combative interests—with or without the involvement of a high school civics class.

Jennifer Pitt told me that she believes accommodation is possible. "Back in 2004, Bennett W. Raley, who was Bush's assistant secretary of the interior, organized a river trip down the Grand Canyon," she said. "The people on the trip were federal brass, state water managers, urban water managers, journalists, and me." The participants included the then head of the Central Arizona Project, who has since retired. He and Pitt had frequently exchanged barbs about the Yuma Desalting Plant, the Ciénega de Santa Clara, and a number of other issues related to the Colorado, but had never met. "Suddenly," she continued, "there we were sitting around the campfire, and I was thinking, 'This is going to be a long river trip.' But we ultimately started talking, and he turned out not to be the Water Buffalo bad guy that some people might have painted him to be, and I wasn't the crazy lunatic fringe environmentalist that some other people might have painted me to be."

One result of their campfire discussions was a yearlong series of meetings about the river which involved a number of interested and often opposing groups, and the meetings were so civilized that the group was eventually able to publish a joint white paper. "That white paper has held the peace to this day," she said. "Now, as things get worse that peace is going to be harder to hold, but just building those relationships was a huge cultural shift." Buzz Thompson told me that Minute 319—part of which can be thought of as having grown out of that white paper and the discussions that led to it—is important in part because it transcends some of the more baffling elements of western water law without discarding them. And if that's the case, the ambiguity of the Law of the River can be thought of as one of its strengths, since it is impossible to be doctrinaire about a doctrine that no one understands.

ACKNOWLEDGMENTS AND
SELECTED REFERENCES

This book began as an article for *The New Yorker*, where I had indispensable help from many people, among them Leo Carey, Jonathan Blitzer, Neima Jahromi, Dorothy Wickenden, and David Remnick. Some of what I know about water I learned on assignment for *Golf Digest*, with help from Mike O'Malley, John Barton, Sue Ellen Powell, Kathy Stachura, and Jerry Tarde. I'm especially grateful to the Environmental Defense Fund and LightHawk for enabling me to look down on the headwaters of the Colorado River from the air (and grateful also to Bradley Udall for suggesting the flight and to Jennifer Pitt for agreeing to go along). I learned a great deal from many helpful people not named earlier, among them Ryan Boggs, Thomas Buschatzke, Joan Clayburgh, Donald Colvin, Chuck Cullom, William Fairbairn, Shelby Futch, Chris Haynes, David Hendrickson, Carly Jerla, Al Kalin, Stephen Koenigsberg, Michael Lacey, David Ludlam, Bronson Mack, Tony McLeod, Bob Muir, Alan Prendergast, Michael Specter, Holly Kirsner Strablizky, Christopher Treese, Jason Walker, Kirby Wynn, and the staff of Western Resource Advocates. Thank you also to my agent, David McCormick; my editor, Courtney Young; Courtney's boss, Geoff Kloske; my wife, Ann Hodgman; our children, John Bailey Owen and Laura Hazard Owen; and our two water-dependent grandchildren, Alice and Hugh O'Keefe, to whom this book is dedicated.

More detailed notes on the text, along with some photographs I took during my travels and links to various documents, are available on the *Where the Water Goes* page at http://www.davidowen.net/. I'll do my best to answer questions and correct errors there, too. Here, chapter by chapter, are a few basic general references, including suggestions for further reading.

CHAPTER 1. THE HEADWATERS

For an aerial view of the entire Colorado River, I highly recommend Pete McBride and Jonathan Waterman's beautiful photo book *The Colorado River: Flowing Through Conflict*. It was published in 2010 and may be hard to find, but it's worth looking for. Also worth looking for is a *National Geographic* map called "Colorado River Basin: Lifeline for an Arid Land." There's also an interactive version on the *National Geographic* website, but you can't spread that one out on the floor. An extremely useful source of technical information about all western water projects, including Colorado-Big Thompson and Fryingpan-Arkansas, is the website of the U.S. Bureau of Reclamation. If you are susceptible to taking freshwater for granted, you should read *Drinking Water: A History*, by James Salzman, published in 2012.

CHAPTER 2. THE LAW OF THE RIVER

For anyone interested in western water law, *Colorado Water Law for Non-Lawyers*, by P. Andrew Jones and Tom Cech, is an excellent place to start. It deals specifically with water law in Colorado, but the basic concepts apply throughout much of the West. The full text of the Colorado River Compact is available online. The interview in which Grady Gammage told his story about unsuccessfully looking up the "Law of the River" was conducted in 2007 as part of an oral history of the Central Arizona Project and is available online on the CAP website, http://www.cap-az.com.

CHAPTER 3. TRIBUTARIES

John Cleveland Osgood and the Colorado Fuel and Iron Company are probably worth a book of their own, as is the history of Crystal Valley. The website of the Redstone Inn will get you started if you are interested in learning more.

CHAPTER 4. GO WEST

Lawrence Wright's *New Yorker* article about mining lithium in Bolivia is "Lithium Dreams"; it appeared in the March 22, 2010, issue. The Los Alamos Scientific Laboratory's informational film about Project Rulison is easy to find on YouTube—where it's listed as "Declassified U.S. Nuclear Test Film #36," although it was never classified—as is the clip of Terry Drinkwater's report on the *CBS Evening News*.

CHAPTER 5. GRAND VALLEY

As with other western water projects, the U.S. Bureau of Reclamation's website is a good source for historical and technical information about the Grand Valley Diversion Dam and its associated infrastructure. *Managing California's Water*, published in 2011 by the Public Policy Institute of California, is easy to find online. The difficulty of turning increases in efficiency into reductions in consumption is the subject of my book *The Conundrum*, which was published in 2011.

CHAPTER 6. SALT, DRY LOTS, AND HOUSEBOATS

There are many accounts of the fight to save Echo Park. A good, brief one is in *John Muir and His Legacy: The American Conservation Movement*, by Stephen Fox, published in 1981—a book that has much else to recommend it.

CHAPTER 7. LEES FERRY

For a week or so, my favorite book of any kind was *Lee's Ferry: Desert River Crossing*, by W. L. Rusho. (The book was published originally as *Desert River Crossing: Historic Lee's Ferry on the Colorado River* in 1975. The reprint edition I own, which I bought in a marina store at Lake Powell, was published in 2003.) The website of the John Doyle Lee Family organization is http://www.johndlee.net/. The best book I've read about the Mountain Meadows Massacre is *Blood of the Prophets: Brigham Young and the Massacre at Mountain Meadows*, by Will Begley, first published in 2002.

CHAPTER 8. BOULDER CANYON PROJECT

My favorite books about my favorite dam are *Colossus: Hoover Dam and the Making of the American Century*, by Michael Hiltzik, published in 2010; *Hoover Dam: An American Adventure*, by Joseph E. Stevens, published in 1988; and *The Story of the Hoover Dam*, edited by C. H. Vivian, published in 1986. That last book contains articles that first appeared in a series of booklets issued, while the dam was under construction, by *Compressed Air Magazine*, an industrial trade publication. (Many big, loud tools used in building the dam ran on compressed air.) The book includes some terrific photographs and illustrations, among many other enticements. I bought my copy in the gift shop at the dam. Edmund Wilson's *New Republic* article about Hoover Dam is anthologized in his book *The American Earthquake*, which was first published in 1958 and is still in print in a paperback edition published in 1979. It's a terrific contemporary look at the effects of the Great Depression.

CHAPTER 9. LAS VEGAS

Details about Las Vegas's xeriscaping rebate program, Water Smart, can be found on the website of the Southern Nevada Water Authority at https://www.snwa.com/. The website of the Las Vegas Springs Preserve and Origen Museum is at https://www.springspreserve.org/.

CHAPTER 10. COLORADO RIVER AQUEDUCT

There are several biographies of William Mulholland, including one by a granddaughter, Catherine Mulholland: *William Mulholland and the Rise of Los Angeles*, published by the University of California Press in 2000. Catherine Mulholland is probably overly sympathetic to and defensive about her subject, but the book is full of interesting material, including a number of cool photographs, and it's a useful counterbalance to books that treat Mulholland as evil incarnate.

CHAPTER 11. CENTRAL ARIZONA PROJECT

Transcripts of the interviews that make up the oral history of the Central Arizona Project can be downloaded from the project's website, at http://www

.cap-az.com. Stewart Udall wrote an early book about the environment, *The Quiet Crisis*, published in 1963, a year after Rachel Carson's *Silent Spring*. It's still a good read.

CHAPTER 12. THE RULE OF CAPTURE

Robert Glennon is also the author of another good book about water: *Unquenchable: America's Water Crisis and What to Do About It*, published in 2009.

CHAPTER 13. BOONDOCKING

Confinement and Ethnicity, the National Park Service's 1999 book about internment camps, resonates today and is worth tracking down. The full text of the Quantification Settlement Agreement for the Colorado River can be found online. You can learn more about military recreational facilities at http://militarycampgrounds.us/.

CHAPTER 14. IMPERIAL VALLEY

There's some interesting historical material about the Imperial Valley in (of all things) *The Story of the Hoover Dam*, that collection of *Compressed Air Magazine* articles mentioned in the note on chapter 8.

CHAPTER 15. THE SALTON SEA

I invited my wife to join me in watching *Plagues & Pleasures on the Salton Sea*, a documentary narrated by John Waters, and doing that was a big mistake because she found the whole thing horrifying and deeply depressing. It will give you a feel for the place, though, as will a companion video compilation, *Past Pleasures at the Salton Sea*. Both are available on CD from Amazon. *Hazard's Toll* is easy to find online. The Redlands Institute has a cool free digital version of the *Salton Sea Atlas* on its website; search for "Salton Sea Digital Atlas."

CHAPTER 16. RECLAMATION

The website of the U.S. Bureau of Reclamation, http://www.usbr.gov/, is vast and terrific. Many potential rabbit holes. A fascinating and potentially endless place to poke around.

CHAPTER 17. THE DELTA

The text of Minute 319 is easy to find online. There's an excellent article about the pulse flow on the website of the Environmental Defense Fund, https://www.edf.org/. Search for "Bringing the River Back to the Sea." A good source of information about salt cedar is the website of the Tamarisk Coalition, http://tamariskcoalition.org/.

CHAPTER 18. WHAT IS TO BE DONE?

I mentioned my visit to the Tampa Bay desalination plant in a *New Yorker* article about Florida sinkholes, "Notes from Underground," published in the March 18, 2013, issue and available online on both the magazine's website and mine, http://www.davidowen.net/.

INDEX